Scientific Discovery
from the Brilliant
to the Bizarre

Also by Len Fisher

The Science of Everyday Life

*(formerly titled How to Dunk a Doughnut:
The Science of Everyday Life)*

Scientific Discovery from the Brilliant to the Bizarre

The Doctor Who Weighed the Soul, and Other True Tales

Len Fisher

ARCADE PUBLISHING

Previously published under the tite *Weighing the Soul: Scientific Discovery from the Brilliant to the Bizarre*

"The Common Cormorant" by Christopher Isherwood is quoted by the kind permission of Don Bachardy and the Christopher Isherwood estate.

Arcade Publishing books may be purchased in bulk at special discounts for sales promotion, corporate gifts, fund-raising, or educational purposes. Special editions can also be created to specifications. For details, contact the Special Sales Department, Arcade Publishing, 307 West 36th Street, 11th Floor, New York, NY 10018 or arcade@ skyhorsepublishing.com.

Arcade Publishing® is a registered trademark of Skyhorse Publishing, Inc.®, a Delaware corporation.

Visit our website at www.arcadepub.com.

10 9 8 7 6 5 4 3 2 1

Library of Congress Cataloging-in-Publication Data is available on file.
ISBN: 978-1-61145-742-1

Printed in the United States

For my father, with fond memories,
and my wife, Wendella, with love

Contents

Preface

All truth passes through three stages: First, it is ridiculed; Second, it is violently opposed; and Third, it is accepted as self-evident.

— Arthur Schopenhauer

This book tells the stories of scientists whose ideas appeared bizarre, peculiar, or downright nutty to their contemporaries but who stuck to their guns through ridicule, oppression, and persecution. Some of their ideas *were* nutty, and most of these ideas (though by no means all!) rapidly became extinct. Other concepts, seemingly every bit as bizarre, passed every test that could be thrown at them and survived to be accepted and used by scientists such as myself as part of our everyday work.

The ideas that scientists now use routinely can still seem ridiculous to people outside science. My wife certainly thought so when she came home one evening to find me riding her bicycle down the road with the wheel nuts removed, explaining to a radio interviewer that the counterintuitive physical laws discovered by Galileo and Newton predicted that the wheels would stay on. Her brief, pungent comment about scientists and their lack of common sense was duly recorded and broadcast on national radio.

My wife was right; science and common sense often don't mix. It's not the scientists' fault; Nature is the principal culprit. Those who proposed bizarre-sounding ideas about its behavior were often forced to do so after recognizing that the accepted wisdom, or "common sense," of their eras was simply insufficient to understand what was going on. Their contemporaries, with a vested interest in maintaining the status quo, were not always as receptive to new ideas as the popular image of the dispassionate, rational scientist would have us believe, and the fates of those who advanced new ideas ranged from the loss of their jobs to the loss of their lives. Their histories belie the popular image of science as an orderly, logical progression. It is more like a procession, with leaders and followers, which is unwillingly forced to change direction each time it comes up against the barrier of a revolutionary new idea. This book traces the route of the procession through the stories of those who forced the changes and shows how many of their ideas, which seemed to be so at odds with the common sense of the time, are now used by scientists to understand and tackle everyday problems. It also reveals the true process of discovery, where the brilliant has often met the bizarre and only the wisdom of hindsight allows us to distinguish between the two. The message is that we need to allow for a certain amount of laughable nuttiness if we are not to lose genuinely original insights and developments. If we can't tell the difference between oddity and insight, then maybe it's wise not to laugh too loud.

A Note on the Approach of This Book

I am a scientist, not a historian, and when I write about scientists from earlier times it is from my perspective as a scientist. In consulting copies of original diaries, papers, and notes, I have often found I was reading about people who thought in the same way that modern scientists do but who happened to be working with a different set of questions and in a different environment of belief about the way in which the world works. I was particularly struck to discover the parallels between their struggles to understand how Nature works and my own efforts (rather less successful) as a child to understand for myself everything from movement, studied by Galileo, to light, space, and time, elucidated by Einstein. I have included some tales from this part of my life, partly to show that thinking like a child isn't necessarily a bad thing when it comes to science, and mainly to show that you don't have to be a genius to understand science — it just needs persistence, and the wish to know.

A Note on the Notes

The notes section in many books is stuffed with boring detail. The notes at the back of this book are different, stuffed with interesting detail and designed to be read independently of the main text. Here you will find unusual, bizarre, and occasionally salacious tidbits, as well as the extra detail and expanded explanations that were too interesting to leave out but which I couldn't fit into the main stories without breaking the flow.

Whether you choose to read the book from the front or the back, its main purpose is still to reveal how science really works, both in the laboratory and in the wider world, and to show that scientists are just as human as anyone else.

Acknowledgments

My wife, Wendella, has been a consistent source of strength and support and has read every chapter from the point of view of a nonscientist. Her amazing ability to spot flaws and obfuscations and to suggest ways of correcting them has added greatly to the ultimate clarity and readability of the book.

My agent, Barbara Levy, has been an enthusiastic proponent from the start, as have my editors, Cal Barksdale and Tom Wharton, whose professional insights have been particularly helpful in maintaining the pace and focus of the story lines. Many of my academic colleagues and other friends have also helped, some in conversation, some by providing material, and others by reading sections of the manuscript with critical and expert eyes that have (I hope) kept me from drifting too far away from the straight and narrow. Among those to whom I owe a particular debt of gratitude are (in alphabetical order) Lindsay Aitkin, Don Bachardy, Peter Barry, Susan Blackmore, Garry Graham, Denis Haydon, Oliver Heavens, Rod Home, Frank James, Jane Maienschein, Jeff Odell, Alan Parker, Larry Principe, Klaus Sander, John Smith, Brian Stableford, Jennifer Woodruff Tait, Phil Vardy, Jeff Watkins, Keith Williams, and Joe Wolfe. Unfortunately, I cannot blame them for any mistakes that remain. It is also the way of such lists that there will be at least one person who has made an important contribution but whose name I have inadvertently omitted. To that person I apologize, with the promise of a drink next time we meet and a correction if this book should run to a second edition.

Scientific Discovery from the Brilliant to the Bizarre

1

Weighing the Soul

I once took part in a science call-in show on an Australian radio station where a builder phoned in claiming to be able to change his weight by the force of his will. He said that he could reduce his weight by a kilogram just by crouching down on his bathroom scale and concentrating, and the reading on the scale proved it.

My caller really believed that he could change his weight. He was not the first to have fallen into this trap. Who among us has not stepped on and off the bathroom scale several times and chosen to believe the lowest weight that was registered? I told him that the reading on his bathroom scale probably depended on his position on the scale platform but congratulated him on performing a real scientific test of his belief. My congratulations were absolutely genuine. That is what real science is all about — checking out beliefs against reality, a process that often involves the accurate measurement of weight.

Scientists, like dieters, are preoccupied with weight. Dieters want to lose it, but scientists want to find it, because it tells us more about Nature than any other single measurement. The founding patron of the Royal Society of London, King Charles II, scoffed at its members for "spending time only in weighing ayre and doing nothing else," but such accurate measurements eventually revealed that air is a very real material composed of a mixture of different gases. It seemed that heat must also be a real material: a tenuous liquid that flowed from hotter to colder places and forced other materials to expand as it entered them. Attempts to weigh heat proved fruitless, however, and forced later scientists to conclude that heat is not a material substance but a form of the immaterial entity that we now call energy.

Heat is not the only immaterial entity that scientists have tried to weigh. Ancient Egyptian tomb paintings show the jackal god Anubis using a set of balance scales to weigh the soul of a recently dead person against a feather. At the beginning of the twentieth century, the American doctor Duncan MacDougall used a modern balance in an actual attempt to weigh the departing soul. MacDougall contended that the soul, if it existed, must have the attributes of a material substance, and his experiments seemed to support his conclusion. This chapter tells the story of those experiments, which were remarkably similar to late-eighteenth-century efforts to weigh heat, and asks why modern scientists find it so easy to believe in a mysterious, weightless entity called energy but find it so difficult to believe in the validity of MacDougall's results. The answers shed light on the true nature of scientific belief and show why scientists can have such a hard time of it when it comes to having their beliefs accepted by other scientists and by the wider community.

I have always been fascinated by the idea that I might have a soul. My early Sunday school teachers inculcated the idea and responded to my incessant questioning as to where the soul might be by saying that it was deep inside me. My earliest scientific experiment was to shine a flashlight down my throat while looking in a mirror to see if I could get a glimpse of my soul, which I imagined would look rather like a Gummi Baby.

I was not the first experimentalist to search for the physical location of the soul. Leonardo da Vinci was denounced as a sorcerer in 1515 for attempting to find the seat of the soul by dissecting the brain, following the belief of the time that the soul, or *senso comune,* had its abode at the center of the head. More recently, the California physiologist Vilayanur Ramachandran and his colleagues have claimed that a specific part of the brain associated with temporal lobe epilepsy is also associated with intense religious experience.

I began to lose my personal belief in the soul when later teachers explained that the soul was something that I could never touch, see, or feel. This worried me, since I could not understand how a soul could touch and affect me if I could not touch and affect it. I did not know that philosophers have struggled for three hundred years with this intractable question, known as the mind-body problem, or that one person had tried to solve it experimentally by attempting to weigh the soul.

The experimenter was Dr. Duncan MacDougall, a physician at a small municipal hospital in Haverhill, Massachusetts. MacDougall, like myself, was not sure whether the soul existed or not, but he was quite convinced that, if it did, it must be a material object that occupied space. He outlined his argument in a letter to his friend Dr. Richard Hodgson, M.D., on November 10, 1901:

> If personal identity (and consciousness and all the attributes of mind and personality) continue to exist after the death of a body, it *must exist as a space-occupying body,* unless the relations here in this world between the conscious ego and space, our notions of space as fixed in our

brain by inheritance and experience are wholly to be set aside and a new set of space relations to consciousness suddenly established, which would be such a breach in the community of nature that I cannot imagine it.

MacDougall's central tenet, that things can only exist in a space that we are able to conceive and visualize, was to take a considerable battering at the hands of Einstein and others a few years later, but even in MacDougall's day it was already obvious from scientific experiment that there were more things in Heaven and Earth than we could dream of, and scientists have repeatedly had to invoke the existence of things outside the range of our imagination just to make sense of the things that we can see, feel, and measure. MacDougall's imagination, though, was limited to the concrete. He believed that the soul, if it existed, must have material form. What he wanted to know was whether that material had weight.

MacDougall does not distinguish in his writings between weight and mass, although the difference has been known since the time of Newton. *Mass* is an intrinsic property of an object and does not depend on where the object is. *Weight* is the downward force that tips a set of scales when the object is in a gravitational field. Objects weigh less on the Moon than they do on Earth because the Moon's gravitational pull on the object is smaller. The force also drops off with distance from the center of Earth (or the Moon), which is one reason records in field events such as the high jump or the shot-put tend to be easier to set at high altitudes. The ultimate place to set such records would be in space, where Earth is so far away that its gravitational attraction is negligible and objects still have the same mass but virtually no weight.

We still don't know what mass is. The most popular the-

Figure 1.1. Astronaut Edwin "Buzz" Aldrin descending from the Lunar Module during the *Apollo XI* moon landing, photographed by Neil Armstrong.
Armstrong later managed a standing jump from the foot of the ladder to a height of nearly six feet despite wearing a heavy space suit that gave him a total *mass* of approximately 135 kg. On the Moon, though, he would have *weighed* one-sixth of that, i.e. 22.5 kg. SOURCE: NASA.

ory is that "empty" space is actually populated by particles called Higgs bosons, which convey the property of mass to any other particles that they come close to. Higgs bosons are tricky things to describe. In 1993, at the suggestion of my Bristol University colleague Professor Sir Michael Berry, the U.K. Minister of Science William Waldegrave challenged physicists to produce a description that would fit on one page. The winning entries, one of which is reproduced in the notes to this chapter, demonstrated the huge gulf between the actuality of Nature and our attempts to understand it using human common sense.

The search for isolated Higgs bosons has so far been fruitless, but it is at least conceivable that there could be a form of matter that does not interact with such bosons and would therefore be massless and weightless, bizarre though that concept may appear. Even in modern-day terms, then, MacDougall's question as to whether the soul material had weight was not unreasonable. His first experimental answer to the question, though, seemed to defy all reason. He describes it in his letter to Dr. Hodgson:

> On the 10th day of last April [1901], my opportunity came. On a Fairbanks Standard platform scales, I had previously arranged a framework of wood, very light; on top of this I placed a cot bed with clothing in such a manner that the beam was not interfered with in any way.
>
> At 5:30 P.M. the patient, a man dying of consumption, was placed on the bed [Macdougall was later at pains to point out that the patient, a young black man, was a fully informed volunteer and that he was not subjected to any additional discomfort]. He lived until 9:10 P.M. During those three hours and forty minutes he lost weight at the rate of an ounce in one hour, the sixtieth part of an ounce in one minute, so that every ten or fifteen minutes I was compelled to shift the sliding weight back upon the beam in order to keep the beam end up against the upper limiting bar, which I wished to do for the sake of making the test of sudden loss all the more

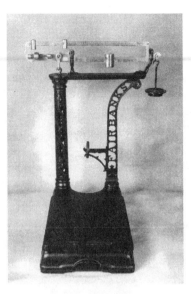

Figure 1.2. Fairbanks Imperial Grocer's Scale manufactured in 1911 and capable of weighing from 1 ounce to 250 pounds.
The platform in the scale illustrated is 12 inches wide by 15 inches deep. MacDougall used a similar type of scale, with the wooden framework and bed fixed to the platform. Unfortunately, he gives no further details of its construction. SOURCE: International Society of Antique Scale Collectors / Edward Konowitz.

marked and decisive, *if such loss should come* [my italics]. This loss of weight . . . was due to evaporation from the nasopharyngeal and bronchopulmonary and buccal mucous membrane accompanying respiration, and also from the evaporation of moisture from cutaneous perspiration.

At 9:08 P.M. my patient being near death, for the last time I sent back the shifting weight on the beam so that for the last ten minutes [*sic*] the beam end was in continuous contact with the upper limiting bar. Suddenly at 9:10 P.M. the patient expired and *exactly simultaneous with the last movement of the respiratory muscles and coincident with the last movement of the facial muscles the beam end dropped to the lower limiting bar and remained there without re-bound as though a weight had been lifted off the bed* [my ital-ics]. Later it took the combined weight of two silver dollars to lift the beam back to actual balance. . . . [T]hese were found together to weigh three-fourths of an ounce.

MacDougall had, it seemed, weighed the departing soul. He wrote to Dr. Hodgson: "Have I discovered [the] soul sub-stance with my weighing machine? I think so, and I mean to verify and re-verify and re-re-verify, if I live long enough."

MacDougall's actions, though, belied his words and re-vealed him as a true scientist. Science is not a matter of dis-covering something and then looking for more and more confirmatory instances. That proves nothing. What good sci-entists do is to try to prove themselves wrong. The more they fail to prove themselves wrong, the more they begin to be-lieve the original idea or observation.

MacDougall went looking for other explanations for the sudden change in weight. With his balance sensitive to a tenth of an ounce, he knew that he had weighed *something*, but he was not at all sure that the something was the soul. His checks, though, did not reveal any other physical explanation. There was no urine loss or bowel movement (although these should not in any case have affected the measurements), and both he and his collaborator Dr. Sproull climbed onto the bed and vig-orously inhaled and exhaled to check whether air loss from the lungs might have an effect. It shouldn't have, and it didn't.

He still wasn't satisfied and tried the experiment with a second patient, who was, like the first, "a man moribund from consumption." The exact time of death of his patient was harder to determine, but the result of the experiment was similar: "Inside of three minutes with all channels of loss closed [i.e. urine, bowel movement, etc.] a loss of one ounce and fifty grains took place."

On May 22 of the following year he was able to write to Dr. Hodgson with the results of four more experiments:

> Since I wrote you last I have had four more experiments on human subjects. In the first of these four there was a loss of half an ounce coincident with death. . . . In the second of the four, the patient dying of diabetic coma, unfortunately our scales were not finely balanced, and although there is a descent of the beam requiring about three-eighths to half an ounce to bring it to the point preceding death, yet I consider this test negative. . . . The third of the four cases shows a distinct drop in the beam registering about three-eighths of an ounce, which could not be accounted for; this occurred exactly simultaneous with death. . . . The fourth case in this series was negative. Unfortunately owing to complications which we could not prevent the patient was but a few minutes on the bed before he died, and whether I had the beam accurately balanced before death or not I cannot be sure.

The results were exciting, but MacDougall was faced with a problem that working scientists face every day of their lives — which results to accept and which to reject? The choice is not a simple one. A wild deviation from an accepted theory or working hypothesis could herald a new discovery, or it could be a misleading and embarrassing artifact. In *How to Dunk a Doughnut,* I described the sad case of the Viennese physicist Paul Ehrenhaft, who tried to measure the charge of the electron by measuring the movement of charged water drops in an electric field. Ehrenhaft performed thousands of experiments and accepted all of his results indiscriminately, so that he could only account for the scatter in his measurements by assuming that the droplets were coated by many electrons, each with a tiny electrical charge. The American Robert Millikan performed similar experiments in America but assumed that there were only a small number of electrons on each drop and robustly rejected results that didn't fit with his hypothesis. As it turned out, Millikan was right and Ehrenhaft was wrong, and the weird, anomalous results that both men had found, but which they had treated very differently, were due not to some new fact of Nature but to dust, droplets sticking together, and other accidental effects.

MacDougall was searching for a new fact of Nature and took the same path that Ehrenhaft later followed by conscientiously describing all of his results, whether they fitted his hypothesis or not. He followed Millikan, though, in looking for reasons to reject results where possible, but unlike Millikan he published his reasons for rejecting them, both in his letters to Hodgson and later when he published the results in the open literature.

By then he had repeated his experiment fifteen times with dogs instead of people. The results were uniformly negative. There was no weight loss on death when men or women were replaced by dogs, and MacDougall cautiously wrote to Hodgson: "If it is definitely proven that there is a distinct loss of weight in the human being not accounted for by known channels of loss, then we have here a physiological difference between the human and the canine at least (and probably between the human and all other forms of life) hitherto unsuspected." He added: "I want first to publish the discovery as a fact in the physiology of death, stripped, as a good friend of mine has said, of its 'psychical significance,' because to insist upon the latter might raise prejudice in the minds of many of our present day scientific men, and prevent repetition of the experiment by others." He was to be sadly disappointed.

Nervous about public reaction, MacDougall kept his results to himself for five years. His only public disclosure was in informal conversation to a group of fellow passengers during a trip to Europe later that year on board the liner *Cestrian*. Despite their encouragement, he refrained from publishing the results of his experiments, fearing ridicule and a community backlash. He had already experienced a backlash from the hospital staff in Haverhill when he was pursuing his experiments. One can imagine the offense that the experiments must have given to the religious and moral susceptibilities of the hospital staff in a small, conservative American township. The objectors attempted to disrupt the experiments. Mac-Dougall mentions, for example, that while attempting to weigh a woman dying of diabetic coma, "there was a good deal of interference by people opposed to our work."

Some years later, an "unauthenticated publication of his attempts" with "the usual distortion of everything that gets in the papers" forced MacDougall's hand. Rather than have distorted newspaper reports as the only public source of his results, he wrote an account of his experiments for the professional journal *American Medicine,* with a duplicate copy in the *American Journal of Psychical Research.* His publication in the latter journal might seem surprising, but research into the possibility of psychic phenomena was much more mainstream then than it is now, and along with anecdotal reports, the journal contained a substantial number of papers reporting rigorous

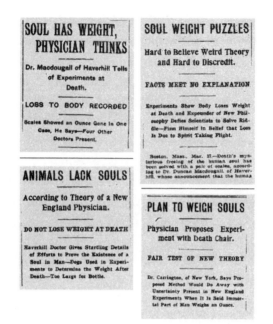

Figure 1.3. Newspaper stories about MacDougall's experiments were generally well balanced and factual (certainly by modern standards!), and it is hard to understand why MacDougall reacted to them as strongly as he did.
SOURCE: *New York Times, Washington Post,* and Haverhill Public Library, Special Collections Department.

experiments designed to test whether particular psychic phenomena really existed and could be demonstrated under controlled conditions that would convince skeptics.

MacDougall maintained a skeptical stance in his published papers. He wrote, for example, "I am aware that a large number of experiments would require to be made before the matter could be proven beyond any possibility of error," and, in a letter written after his results became public, "I am well aware that these few experiments do not prove the matter any more than a few swallows make a summer." Nevertheless, he felt driven by his own measurements into a corner, and he wrote privately, "Is it the soul substance? How else shall we explain it?"

MacDougall's account of his experiments makes fascinating reading today. One cannot help but be struck by the care with which the work was done and the detail in which it is presented. A present-day journal referee would be hard-pressed to find a reason for rejecting it and would probably say that the conclusions seemed incredible but that they could find no basic flaw in the methods or reasoning. The paper would then be published (as many are) in the expectation that other scientists would try to repeat the experiments and either find the flaw or reinforce the conclusion. That is the way science works; at least, it is the way that it is supposed to work.

MacDougall's measurements seemed to show that the soul weighed as much as a slice of bread. It was an extraordinary claim, and in Carl Sagan's pithy phrase, "Extraordinary claims require extraordinary proof." Some such claims, like the supposed discovery of a new form of water in 1970 or the possible production of cheap, safe energy from "cold fusion" in the late 1980s, led to bursts of scientific activity that eventually proved them wrong. Others, like the idea that huge polymer molecules such as proteins could fold up spontaneously to produce regular structures or that light sometimes behaves like a wave and sometimes like a particle, have proved to be right. Still others, such as the claim that the arrangement of the stars as viewed from our planet somehow affects our behavior and our future, have repeatedly been shown to have no demonstrable foundation yet still claim large numbers of adherents.

MacDougall's methods appeared to be perfectly sound, but his result was so extraordinary that no one has yet attempted to repeat his experiments — at least, not with people. Similar experiments, though, were performed in the 1930s on some unfortunate mice. The perpetrator of these experiments was the Los Angeles schoolteacher H. LaVerne Twining, who was one of the early aviation and radio pioneers. His book *Wireless Telegraphy and High Frequency Electricity,* which he self-published in 1909, is now a rare gem, and he was the first member of the Aero Club of California to build a monoplane that flew. Perhaps ascending to those dizzy heights stimulated his interest in heavenly things, because by 1915 he had already written a book entitled *The Physical Theory of the Soul,* although he did not publish his experiments until more than twenty years later.

To perform his experiments, Twining set up a balance with a glass beaker on either side. Each beaker contained a live mouse, and on the balance pan beside each beaker was a solid lump of potassium cyanide. With the whole arrangement exactly poised, Twining lifted one of the lumps of cyanide with a pair of tweezers and put it into the beaker on the same balance pan. Within thirty seconds the mouse in the beaker died. When the mouse died, the balance arm dropped! Mice, it seemed, also had weighty souls.

The mouse that died might, of course, have been a particularly saintly mouse. Twining, however, like MacDougall before him, was more inclined to seek a physical explanation. Unlike MacDougall, he had another experiment up his sleeve — a different form of death. Instead of using cyanide, he sealed another unlucky mouse in a glass tube so that the poor animal died of asphyxiation (a cruel experiment that would never be condoned by modern ethics committees). This time, there was no weight loss when the animal died.

Twining concluded that a dying mouse somehow loses moisture rapidly at the point of death and that the moisture remains trapped when death takes place in an enclosed chamber. That would seem to answer the question except for the fact that Twining didn't actually test for moisture. One might equally conclude from the evidence that mice have substantial souls, but that those souls can't get through glass.

In fact, the idea of rapid moisture loss doesn't really explain either Twining's or MacDougall's results. It is just too hard to see how the death of an animal, human or otherwise, could result in such a rapid expulsion of moisture. Could there be another physical explanation? Or had MacDougall, after all, really weighed the soul?

MacDougall was unable to conceive of an alternative physical explanation, but I came across an interesting possibility while researching early ideas about the nature of heat and electricity. These days we view them as different forms of energy that can be turned into each other, as when we use electricity to heat a radiator or when we burn fuel to drive an electric dynamo, and that can be used to do work, as when we drive an electrically powered golf cart or ride in an old-fashioned coal-burning steam engine. The everyday language that we use to describe heat and electricity, however, still reflects the early pictures that they were real fluids that "flowed" respectively from hot to cold places or along wires. Such literal pictures held considerable sway even up until recent times.

The images of heat and electricity as real, material fluids held sway because they fitted with common sense. Materials increase in volume when they get hotter, for example, which is just what you would expect if they had to expand to accommodate the added fluid. The fluid was even given a name — *caloric.* Belief in the reality of caloric was so widespread until the middle of the nineteenth century that if you look up *heat* in an encyclopedia of the time, the entry will say "See *caloric.*" The name was popularized by one of the best scientists of the period, the French aristocrat Antoine Lavoisier, who listed caloric as one of the twenty-four elements in his *Traité élémentaire de chimie* and who later lost his head in the French Revolution (though not for the proposal of caloric).

There was plenty of evidence that caloric was a real material, not just from common sense but also for more arcane scientific reasons, such as the fact that it permitted a correct prediction of the ratio of the specific heats of liquids and gases, a matter of great interest to scientists at the time. Everything seemed to point to the reality of caloric, but no one had isolated it, no one had seen it, and no one had managed to weigh it. By the end of the eighteenth century, there were some who were beginning to doubt it existed at all. One of these was the redoubtable Count Rumford, who had started life as plain Benjamin Thompson in America before maneuvering his way into the high politics of Europe, eventually reorganizing the German army, and being raised to the rank of a count for his trouble. His experiments with the boring of brass cannons in Munich had showed him that he could generate as much heat as he liked just by keeping the boring process going. This convinced him that heat could not be a real material, because the supply of the material would surely have run out at some stage. To prove his point, he designed an experiment that was remarkably similar to MacDougall's later attempts to weigh the soul.

The experimental plan was ingenious. Just as the soul is expected to leave the body at death, so heat is expected to leave a liquid when it freezes. Rumford thus decided to check whether there was any change in the weight of a container of water as heat was extracted from it by refrigeration to turn the water into ice. For this he needed a highly accurate balance, which was beyond his means, but Rumford used his influence in Bavarian court circles to borrow one "belonging to *His most Serene Highness* [Karl Theodor] *the* ELECTOR PALATINE DUKE *of* BAVARIA," patron of Mozart and others. His refrigerator was the Bavarian winter, and when "the cold was sufficiently intense for my purpose" he set up his balance (similar in design to that of the ancient Egyptians!) with some 250 milliliters of water in a sealed flask on one side and "an equal weight of weak spirit of wine" on the other. He kept the approximately balanced assembly in a heated room until he could be sure that it had come to thermal equilibrium and brought the weights into exact balance by hanging a piece of fine silver wire on one balance arm. Then, holding his breath, he moved the whole assembly to a quiet unheated room where the temperature was −2°C and, "going out of the room, and locking the door after me, I suffered the bottles to remain forty-eight hours, undisturbed, in this cold situation."

At this temperature, he expected the water to freeze but the spirit of wine to remain liquid, because alcohol lowers the freezing point of water (a fact that allows me to keep my gin and vodka in the freezer compartment of the refrigerator). His expectations were right — when he carefully opened the door of the room after two days, the water had turned to ice but the spirit of wine had remained liquid. To his chagrin, however, the balance arm had tilted from its original horizontal position. Heat, it seemed, did have weight!

That wasn't the only unpleasant surprise. The tilt of the balance arm was in the wrong direction; the side of the balance carrying the water fell instead of rising. Heat appeared to have *negative* weight!

Like MacDougall, Rumford's first inclination was to repeat the experiment. He did so, and obtained a similar result. If he had continued to repeat the experiment over and over in the same way, he would have gotten the same result again and again and would presumably have been forced to accept his unpalatable conclusion. This is what happened to MacDougall in his attempts to weigh the soul. But Rumford knew that the most convincing results are those that have been obtained in several different ways, and he tried a different approach. After checking the balance itself — to make sure that the two balance arms did not shrink by different amounts as they cooled — he replaced the spirit of wine in one bottle with mercury (a better conductor of heat that would come more rapidly to thermal equilibrium) and repeated his experiment. To his intense pleasure, the balance remained "in the most perfect equilibrium" as the water froze. Finally, he had shown that heat had no weight within the sensitivity of his apparatus.

So why had Rumford's earlier experiments appeared to have shown that heat had negative weight? The answer, which also bears on the interpretation of MacDougall's and Twining's experiments, was that the two sides of the balance were at different temperatures. When water freezes, the water and the ice that is formed from it both remain at a temperature of zero degrees centigrade until all of the water has frozen. The flask on the other side of the balance, though, would have kept on cooling. The difference in temperature between the two sides would have caused air movement, just as it does on a larger scale when we put a convector radiator in a cold room, or on a still larger scale with the air movements over warm land and cold water that control our weather.

Precision balances such as that used by Rumford are particularly susceptible to air currents, and the readings have no value until the air movement has stopped. Convection currents certainly affected Rumford's early experiments, and it seems possible that they could also have arisen in MacDougall's and

Twining's experiments when the bodies that they were studying cooled down upon death. Rumford eventually overcame
the problem by creating experimental conditions where thermal equilibrium could be established, but neither MacDougall
nor Twining allowed for the possibility of convection currents.
Their attempts to weigh the soul would both have been susceptible to such air movement (except for the experiment
where the dying mouse was confined to a sealed tube, which is
the experiment where no weight change was observed). It is a
difficult effect to allow for, as Rumford found, but one that
must be taken into account before MacDougall's experiments
could possibly be interpreted as having weighed the soul. The
null result with dogs admittedly poses a difficulty for the "convection" explanation, but the dogs were covered with insulating hair, which may have affected the results. In any event,
convection currents must be eliminated before any future experiment similar to MacDougall's could possibly be interpreted
as weighing the soul.

There is a lot more to this little story. Even though
Rumford eventually got his experiments right, they still did
not convince believers in caloric, who simply assumed that
caloric must therefore be weightless, a view that continued to
have weight for a further fifty years. Likewise, refutation of
MacDougall's experiments is unlikely to affect believers in
the soul, most of whom in any case think of the soul as an insubstantial entity rather than as one with physical form. There
is *always* another possible conclusion for believers in a particular theory. Facts can always be rearranged to fit a theory, although the conflict between belief and reality can become
well-nigh insurmountable, as with the letter arriving at the
offices of the Flat Earth Society announcing that the recipi-
ent had won a round-the-world trip.

The reception of both Rumford's and MacDougall's results demonstrates the difficulty of performing a true *experimentum crucis*. It also points up an important fact of life, which is that belief in science or religion is a complex business, based on what the believer finds convincing. Scientific beliefs, like religious beliefs, often start life as an article of faith held by one or two people. The difference is that religious beliefs, such as belief in the soul, often continue to be held because the holders cannot make sense of the world without them, while scientific beliefs only continue to be held if their predictions conform to observable reality. This does not make scientific beliefs any more valid — they are just buttressed in a way that is more concrete and convincing to many people.

Rumford's belief was that heat is "motion," which was a proposal that my first physics teacher tried to prove to me by pointing out that my hands got hotter when I rubbed them together. Rumford's idea of heat as motion, like Lavoisier's idea of heat as the fluid "caloric," was based on common sense — if heat wasn't a material itself, then what else could it be except movement within a material? He was half right — when I rub my hands together, the molecules in the skin are made to vibrate faster. It took the combined efforts of many scientists, however, to discover that Nature's actuality is far beyond the limits of our direct sensory experience, and to arrive at the belief that heat is just one form of "energy," a mysterious, insubstantial entity that, like the soul, we will never be able to touch, see, or feel but only experience in its different manifestations.

There are many forms of energy — light, electricity, and heat, to name just three. Scientists believe that these are all forms of the same basic thing because they can be converted backward and forward into each other in a quantitative way. For example, if I shine light onto a photocell (such as those used to power some roadside signs), the light is transformed into electricity. If I then use the electricity to drive a laser, the amount of light that I can get from the laser will be the same as that which I originally shone onto the photocell (with an allowance for loss of some energy as heat during the two processes).

If that were all that there was to it, then the idea of "energy" would be no more than an academic plaything. The real point about energy is that it can be made to do work, i.e., move things. In fact, that is the definition of energy — anything that can be made to do work. Heat and electricity can be used to drive engines, and even light can be used to spin a small fan called a Crookes' radiometer. One form of movement can also be used to drive another, as in Heath Robinson's marvelous drawing of a bicycle action being used to drive a potato peeler, so movement is also a form of energy, called kinetic energy.

Energy is one of those words that scientists have stolen from our everyday vocabulary, but luckily the scientific definition coincides with the ordinary meaning — the more energy you have, the more work you can do. The discovery that energy comes in many different, interconvertible forms also gives us the opportunity to store and transport it in one form and then convert it and use it in another form. Unfortunately, every conversion or transport process so far discovered involves waste. When a fluorescent tube is placed close to a high-voltage power line, for example, the tube begins to glow as it picks up the energy given off and wasted as low-frequency radio waves.

Even some of the energy stored in a battery is wasted when the battery is used — just run your car lights off the battery for a few minutes, and then feel how hot the battery has become. That heat is wasted energy. The problems of wasted energy and of finding a convenient, lightweight method of storing and transporting the energy — which we could produce in remote regions by environmentally friendly methods such as wind turbines or solar panels — are among the most urgent practical problems facing us today. One of my hopes for this book is that it might stimulate more people to become involved in this or in one of the many other everyday, practical problems to which science is essential for the solution.

Figure 1.4. Unpowered fluorescent tubes shaped to resemble human brains absorb sufficient energy from the electromagnetic field to glow when placed under high-voltage power lines near Bristol, U.K.
The message is not necessarily a frightening one. It requires very little power to cause a fluorescent tube to glow, and any effect on real human brains of this level of power, and at these frequencies, is still to be convincingly demonstrated. SOURCE: Richard Box, 2003 Artist-in-Residence, School of Physics, University of Bristol, U.K.

2

Making a Move

It filled Galileo with mirth
To watch his two rocks fall to Earth.
He gladly proclaimed,
"Their rates are the same,
And quite independent of girth!"

— Entry in American Physical Society limerick competition

Galileo established the modern science of motion against the violent opposition of his contemporaries, who still believed in Aristotle's ancient ideas that heavier objects fall faster than light ones and that objects only move while they are being pushed or pulled. Galileo's idea was that objects would keep moving at the same speed even when the pushing stopped. His opponents thought the idea ridiculous, but it has been confirmed many times. A very senior British scientist once told me with glee about how a group of his colleagues had inadvertently confirmed it in spectacular fashion while moving a large, heavy magnet mounted on a trolley. They pushed and perspired to roll the trolley from the side of the laboratory to its new position in the center, and then stopped pushing. Unfortunately for them, the trolley kept on rolling, eventually to disappear with a crash through the opposite wall, showing conclusively that Galileo was right.

Galileo was often right, and when he was appointed as a junior lecturer at the University of Pisa he was not afraid to show his contempt for his Aristotelian colleagues. His caustic comments went down well with his students, but it was his colleagues who held the power, and they saw to it that his lectureship was not renewed after the statutory three years. They pursued him with innuendo and criticism even after he moved to the more liberal independent state of Venice, but their vendetta could not stop his continuing investigations into the true nature of motion. The story of how his investigations into the motion of heavenly bodies led to conflict with the Inquisition and imprisonment by Pope Urban VIII is well known. This chapter tells the less well known story of how Galileo used his time of imprisonment to set down the results of his lifetime of discoveries about mechanics and motion in the *Dialogues Concerning Two New Sciences* and of how the manuscript was smuggled out of Italy to become the best-selling book that set science on its modern path.

The reliquiae of Galileo's discoveries are lodged in the Italian History of Science Museum, hidden away behind the Uffizi Gallery in Florence. I first came across them when I had tired of the long lines at the Uffizi and walked away from the artworks to look at the equally beautiful old astronomical instruments next door. Here I came across Room IV, devoted to Galileo, and found myself alone with the relics of the man who set science on its modern, skeptical, questioning path, and who provided many of the practical means for following that path. Here was a mechanical calculator, a machine for raising water, a device for calculating longitude while at sea, a "thermoscope," and a magnetic device for moving large weights — precursors, respectively, of the computer, the hydraulic pump, modern navigational aids, the thermometer, and the magnetic cranes that are used to lift and drop car bodies in wrecking yards.

Here also was a lens from the telescope with which Galileo later discovered the moons of Jupiter, a discovery that showed there had to be more than one center of motion in the universe, undermining Aristotelian cosmology and leading Galileo to question whether the earth occupied the unique position claimed for it by the Catholic Church. Another, macabre relic was the mummified middle finger of Galileo's right hand, tastefully mounted on a small alabaster base and pointing upward to the sky. It seemed to me that he was raising it to the establishment.

We see Galileo now as a rebel against the establishment, but the truth is that he always wanted to be a member of that establishment. When he began his career as a mathematician, his first concern was to obtain a secure academic appointment at an established university, and he followed a course that today's Ph.D. students still follow — he looked for a safe problem that would produce guaranteed results and establish his reputation. The problem that he chose was *balance,* and over several years as a private mathematics tutor without a tenured academic job he used all his spare time to calculate the centers of balance of differently shaped objects. He also analyzed Archimedes' method for using a balance to find the relative densities of different substances and described his results in his first book, *La Bilancetta* (*The Little Balance*). It is very readable, even now. I wish it had been a textbook when I began to study physics in college, because it explains the principles of hydrostatics much better than my lecturers ever did.

Galileo circulated his book in manuscript form, and it became an instant success. Together with his other calculations, it established him as a mathematician to be reckoned with. The academic establishment welcomed him with open arms and asked him to apply his mathematical ability to what was to them a far more important problem — the calculation of the exact dimensions and location of Hell, as described in Dante's *Inferno*. Galileo took his assigned task seriously. He began by assuming that Earth was the center of the universe, and that Dante's seven levels of Hell were all below Earth's surface, covered by a cap that was a part of that surface. Over the course of two lectures to the Florentine Academy he used mathematical arguments to demonstrate that Hell must have a shape like an ice-cream cone, with the point at the center of Earth and the circular boundary of the base passing through Jerusalem.

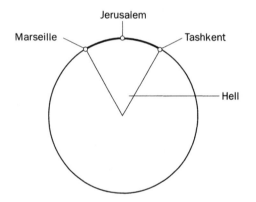

Figure 2.1. Galileo's model of Dante's Hell as a conical sector in Earth.

Galileo's main problem was the strength of the vaulted roof that covered Hell. According to his calculations it covered a span of around 3,000 miles, but it was less than 330 miles thick, and it seemed that it should collapse into the void below under its own weight. Galileo argued that this would not happen because the vaulted roof was just a scaled-up version of

Brunelleschi's famous dome of the Florence Cathedral, which supports its own weight perfectly well. His argument convinced his aristocratic audience, and he was rewarded with a lectureship in mathematics at the University of Pisa, where he soon realized that he had gotten the mathematics of Hell terribly wrong. The domed roof he had proposed was bound to fail, because a structure that is strong on a small scale will collapse under its own weight when scaled up to a much larger size. Galileo never admitted his mistake publicly, but it stimulated him to a lifetime of largely secret work on how the mechanical strength of objects changes as their dimensions change — a subject that is now known as scaling theory.

Galileo's work on the strength of materials was one of the two "New Sciences" that he was later to write about. The other was motion, where he quickly went on the attack against his conservative Aristotelian colleagues. His first line of attack was personal. He jeered at their wearing the traditional academic dress (the toga) in a 301-line poem "Against the Donning of the Gown," claiming that such uniforms concealed a person's true character, prevented men and women from appraising each other's attributes, and even made it difficult for professors to visit a brothel without being recognized. The serious target for his attack, though, was Aristotle, and especially Aristotle's Law of Falling Bodies. He made fun of Aristotle's notion, as he had made fun of the toga, by taking it to extremes:

> [W]ho will ever believe that if, for example, two lead balls, one a hundred times as large as the other, are let fall from the sphere of the moon, and if the larger comes down to the earth in one hour, the smaller will require one hundred hours for its motion? . . . Or, again, if a very large piece of wood and a small piece of the same wood,

the large piece being a hundred times the size of the small one, begin to rise from the bottom of the sea at the same time, who would ever say that the large piece would rise to the surface of the water a hundred times more swiftly?

One of Galileo's pieces of ammunition against Aristotle was a very clever argument that he later summarized in *Dialogues Concerning Two New Sciences,* in which he uses a Socratic dialogue approach, making his own points under the assumed name of Salviati, and having them received with open-mouthed admiration by a dull-witted character called Simplicio (the alter ego of his opponents):

> SALVIATI: If then we take two bodies whose natural speeds are different, it is clear that on uniting the two, the more rapid one will be partly retarded by the slower, and the slower will be somewhat hastened by the swifter. Do you not agree with me in this opinion?
> SIMPLICIO: You are unquestionably right.
> SALVIATI: But if this is true, and if a large stone moves with a speed of, say, eight while a smaller moves with a speed of four, then when they are united, the system will move with a speed less than eight; but the two stones when tied together make a stone larger than that which before moved with a speed of eight. Hence the heavier body moves with less speed than the lighter; an effect which is contrary to your supposition. Thus you see how, from your assumption that the heavier body moves more rapidly than the lighter one, I infer that the heavier body moves more slowly.

Wonderful! The combined body would be heavier than either, and hence have a greater "natural speed," but the process of combination should logically lead to a body with an intermediate "natural speed." The only way to avoid the paradox is to assume that there is no such thing as a "natural speed" that depends on the size of a body, and that all bodies must instead fall at the same speed. The method of "proof by contradiction" that Galileo uses here was invented by Euclid, and used by him to prove that there must be an infinity of prime numbers. When I first came across it while studying mathematics in college, I thought it was the most beautiful logical idea I had ever seen in my life (I still do). No one before Galileo, though, had dared to apply it to a purely physical problem.

But Galileo used more than logic to make his point. He was the first scientist to challenge logic with experiment, and to take experiment as the truer test of an idea. His experiment was simply to take two objects of different weight and drop them together from a height. It seems amazing that no one in the two thousand years since Aristotle had actually tried this, but Galileo did. He recalled the experiment in his final book:

> But I, . . . who have made the test, can assure you that a cannon ball weighing one or two hundred pounds, or even more, will not reach the ground by as much as a span ahead of a musket ball weighing only half a pound, provided both are dropped from a height of 200 cubits.

A cubit, originally defined as the length of a forearm, was settled in Galileo's day as 18 inches, so Galileo was talking about a drop from a height of around 300 feet, or 92 meters. The Leaning Tower of Pisa is 185 feet, or 56 meters, high, so it is unlikely that this was the location of the experiment, despite the popular story. Historians have argued that this story was invented by his biographer, Vincenzo Viviani, the young man who helped Galileo to write *Dialogues Concerning Two New*

Sciences when he was old, imprisoned, and blind, but Galileo's own account shows that he performed a similar experiment somewhere. The Campanile di San Marco in Venice is a candidate, at a height of 99 meters, but even in those days St. Mark's Square was a favorite spot for tourists, who would not have relished the prospect of cannonballs plummeting from the tower above them. Perhaps this is why Galileo never revealed the exact location of his experiment.

My own suspicion is that he tried the experiment more than once, especially since his first try produced a result that was very peculiar indeed — he found that the lighter ball drew ahead at the start of the fall, but that the heavier ball eventually overtook and passed it! When a group of students at Rice University replicated Galileo's original experiment by dropping equal-sized spherical balls of wood and iron from a height, they discovered the same effect. The explanation, revealed by high-speed photography, was that the experimenters couldn't help gripping the heavier iron balls more firmly at the start, so that they took longer to be released from the hand than did the wooden balls. The movement of the wooden balls was more affected by air resistance, though, so the iron balls eventually caught up and passed them.

Air resistance can dramatically affect the speeds of falling bodies — just think of the function of a parachute, which slows a fall by its resistance to motion through the air. My father, an aircraft instrument inspector, taught me about the effects of air resistance when I challenged his assertion that all objects fall at the same speed, arguing that a piece of wood falls faster than a feather. His answer was to take a flat wooden board, place a feather on top of it, and drop it from shoulder height. The feather, protected from air resistance by the wood, dropped with it at the same speed, and I learned my first lesson in science — look before you leap, or drop.

If Galileo had looked before he leaped, he might not have gotten into the trouble in which he found himself. Instead of being politically quiet about his discoveries, he was openly scathing about his colleagues who continued to hold Aristotelian views. "Men are like wine flasks," he once said to a group of students. "[L]ook at . . . bottles with the handsome labels. When you sample them, they are full of air or perfume or rouge. These are bottles fit only to pee into!" It is hardly surprising that his fellow professors were disinclined to look objectively at his criticisms of Aristotle's theories of motion, which he put into a manuscript called *De Motu (On Motion)*. Instead, they vented their wrath by ensuring that his three-year contract at the University of Pisa was not renewed.

Having leaped, though, Galileo landed on his feet. At the age of twenty-eight he was head-hunted for a job at the University of Padua, the alma mater of Dante, in the independent state of Venice, at three times the salary of his former post. Here he remained for another eighteen years, saying later that this had been the happiest time of his life, and it was here that he was able to bring his studies on motion to full fruition.

His first concern was to work out the true law of falling bodies, and he correctly saw accurate timing of the fall of the balls as the key. His ingenious solution was to reduce the speed of the balls by rolling them down an inclined plane instead of dropping them from a height. He must have built many inclined planes over the years that it took him to perfect the experiment, but none have survived, and the model that may now be seen in Room IV of the Florence Science Museum represents a "best guess" based on Galileo's descriptions. With the aid of such planes, he finally discovered the true law of falling bodies, which is that the distance fallen is proportional to the square of the time involved, and totally independent of the weight or density. So, no matter how much you weigh, if you fall off a high ladder, you will have fallen four times as far after two seconds as you do after one second. The message of Galileo's mathematics is to grab hold of a rung of the ladder earlier, rather than later, in your fall.

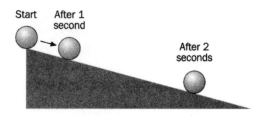

Figure 2.2. The principle of the inclined plane.

Galileo's career of scientific discovery was not the only thing that took off. He had a taste for the high life, and the Venice of 1592 was just the place to find it. The high life came in the form of a sybaritic lifestyle hosted by the Venetian playboy Gianfrancesco Sagredo at his palatial villa on the bank of the Brenta canal. The Villa Sagredo, built on top of a Roman ruin, is now famous for its gardens, but in Galileo's day it was the scene of wild parties that earned it considerable notoriety. Here Galileo would spend many of his weekends, and it was here that he met Marina Gamba, with whom he had three children, although they never married, nor even lived in the same house during the twelve years of their relationship. He also spent much time with his friend Filippo Salviati, with whom he shared a taste for burlesque poetry and low comedy. It was at Salviati's magnificent villa overlooking the river Arno near Florence that Galileo used his newly invented telescope to observe sunspots in 1612. These were a first hint that religious arguments about the perfection of the heavens did not square with experimental observation.

Galileo did not invent the telescope. It was a Dutch invention, which the Dutch authorities tried to keep quiet because of its obvious military potential. It didn't stay secret for long because, as with most scientific secrets, there were too many people thinking along parallel lines. Within a year word had gotten around that there was a device that allowed its user to view distant objects. When Galileo heard about it in the spring of 1609, it didn't take him long to work out how it must have been done and to make his own three-power telescope from spectacle lenses. By August he had taught himself the art of lens grinding and had produced an eight-power instrument that he presented to the Venetian senate, which rewarded him with life tenure at the University of Padua and a doubling of his salary. At the age of forty-five, the world was his oyster. But then, in December, with a telescope that was now capable of twenty-fold magnification, he pointed his instrument at the skies and found that the Moon's surface, then thought to be the embodiment of heavenly perfection, was rough and uneven. In the following January he discovered four moons revolving around Jupiter, and found that his telescope revealed many more stars than were visible to the naked eye. He was on his way to heresy.

By March, Galileo, working at a furious pace, had written up his discoveries in a little book called *Sidereus Nuncius (The Sidereal Messenger),* perhaps the most dramatic science book ever published. With political acumen, he dedicated the book to the powerful grand duke of his native Tuscany, Cosimo II de Medici, and named the moons of Jupiter after the Medici family; Medici I, II, III, and IV, names that they were to retain for a couple of centuries. His reward from Cosimo de Medici was an appointment as mathematician and philosopher to the Tuscan court, and in October of 1610 Galileo returned in triumph to his native land to live the life of a gentleman. Here is the account of his discovery of the "Medici planets" in his own words:

I should disclose and publish to the world the occasion of discovering and observing four Planets, never seen from the beginning of the world up to our own times, their positions, and the observations made during the last two months about their movements and their changes of magnitude; and I summon all astronomers to apply themselves to examine and determine their periodic times, which it has not been permitted me to achieve up to this day. . . . On the 7th day of January in the present year, 1610, in the first hour of the following night, when I was viewing the constellations of the heavens through a telescope, the planet Jupiter presented itself to my view, and as I had prepared for myself a very excellent instrument, I noticed a circumstance which I had never been able to notice before, namely that three little stars, small but very bright, were near the planet; and although I believed them to belong to a number of the fixed stars, yet they made me somewhat wonder, because they seemed to be arranged exactly in a straight line, parallel to the ecliptic, and to be brighter than the rest of the stars, equal to them in magnitude. . . . When on January 8th, led by some fatality, I turned again to look at the same part of the heavens, I found a very different state of things, for there were three little stars all west of Jupiter, and nearer together than on the previous night.

I therefore concluded, and decided unhesitatingly, that there are three stars in the heavens moving about

Jupiter, as Venus and Mercury around the Sun; which was at length established as clear as daylight by numerous other subsequent observations. These observations also established that there are not only three, but four, erratic sidereal bodies performing their revolutions around Jupiter.

Galileo's studies of motion had borne forbidden fruit. They showed that not all heavenly bodies revolve around Earth as a center and began to convince him that the Copernican picture of a Sun-centered solar system was probably right. But not all people were convinced by his observations. Galileo's telescope was undoubtedly primitive, and it took a practiced eye to observe what Galileo said he had seen. Many people who tried to look through Galileo's telescope had problems similar to those of James Thurber when he tried to see plant cells down a microscope. "I see what looks like a lot of milk," he told his instructor, whose patience was coming to an end:

"We'll try it," the professor said to me, grimly, "with every adjustment of the microscope known to man. As God is my witness, I'll arrange this glass so that you see cells through it or I'll give up teaching. In twenty-two years of botany, I —" He cut off abruptly for he was beginning to quiver all over. . . . So we tried it with every adjustment of the microscope known to man. With . . . one of them . . . I saw, to my pleasure and amazement, a variegated constellation of flecks, spots, and dots. These I hastily drew. The instructor, noting my activity, came back from an adjoining desk, a smile on his lips and his eyebrows high in hope. "What's that?" he demanded, with a hint of a squeal in his voice. "That's what I saw," I said. "You didn't, you didn't, you didn't!" he screamed. . . . "That's your eye! . . . You've fixed the lens so that it reflects! You've drawn your eye!"

Some of Galileo's contemporaries avoided such potential problems by simply refusing to look through his telescope. One of these was Professor Guilo Libri, a philosopher at the University of Pisa who had been a persistent critic of Galileo. Invited to view the heavens through Galileo's telescope, he replied that there was no need to do so, because he knew the truth already. When Libri died a few months later, Galileo tartly commented that, although Libri would not take a look at celestial objects while on earth, perhaps he would take a view of them on his way to Heaven.

When Galileo turned his telescope to Venus, he observed that it has phases, as the Moon does. This was proof that Venus traveled around the Sun, as the Moon travels around Earth. When he turned his telescope to the Sun, and observed dirty black moving spots on its surface, he was moved to argue that these were imperfections in the surface of the Sun, and not satellites crossing in front of it as the Jesuit mathematician Christoph Scheiner had claimed in his book *Tres Epistolae de Maculis Solaribus (Letters on Solar Spots)*. The argument raged back and forth, and eventually Galileo was provoked to claim on paper that the Copernican heliocentric theory was the only one that fitted his telescopic observations:

> And perhaps this planet [Saturn] also, no less than horned Venus, harmonizes admirably with the great Copernican system, to the universal revelation of which doctrine propitious breezes are now seen to be directed towards us, leaving little fear of clouds or cross-winds.

There were clouds on the horizon for Galileo, though. His acid tongue and his propensity for challenging statements had made him many enemies, and his move from independent Venice back to Tuscany had left him a target for the wrath of the pope. When, at the age of sixty-eight, he finally published his discoveries about the movement of heavenly bodies in a book entitled *Dialogue Concerning the Two Chief World Systems,* the pope was furious. Thirty-two years earlier the Italian philosopher Giordano Bruno had been burned at the stake merely for suggesting that the universe might be infinite, in opposition to contemporary interpretation of Holy Writ. Given this relatively recent precedent, it might seem that Galileo was foolhardy to challenge the church's authority by claiming that Earth was not the center of the universe. In mitigation it must be said that Galileo, emboldened by his social position to write the book, had nevertheless been careful to submit it to the Inquisition's censor, who had given him permission to publish it so long as the idea of a heliocentric solar system was advanced as a "hypothesis." He had even directed Galileo to incorporate the opinion of the pope on the matter. Galileo did, but he was unwise, to say the least, to put Pope Urban VIII's opinion into the mouth of the dull-witted Simplicio.

Galileo was asking for it, and he got it. In a story that is too well known to bear repeating in detail, the pope called Galileo to Rome to face a court of ten cardinals. The pope himself stayed away, but the members of the court knew what was expected of them, and Galileo was threatened with torture and forced to "abjure, curse and detest" his heretical view that Earth goes around the Sun. His book was placed on the infamous *Index Librorum Prohibitorum* — the *Index of Banned Books* — and he was placed under house arrest for the rest of his life.

It was here that Galileo returned to the basic studies on the strength of materials and the movement of earthly bodies that had occupied him for most of his life. He had already drawn his conclusions. Now it was a matter of putting them on paper, which he did in the manner of a Socratic discussion between three characters: Simplicio (representing Galileo's Aristotelian enemies), Salviati (representing Galileo), and Sagredo, a wise nonpartisan observer named after Galileo's Paduan friend. These three appear as nice, friendly characters, chatting about practical problems and their scientific solutions in very much the same way that my friends and I do when we meet in the coffee room of the Physics Department at Bristol University each morning to chat about our latest results.

Galileo's house arrest was not entirely onerous. He spent the first part of it as an honored guest in the home of the Tuscan ambassador before moving to the residence of the archbishop of Siena, where he was given facilities to start writing *Dialogues Concerning Two New Sciences,* which he continued to write after finally being allowed to return to his own villa in Arcetri. The conversation in it between the three S's takes place over four days. For the first two of those days Galileo returns to the problem of the vaulted roof covering the cone of Hell. He doesn't mention his original calculation — the memory of his failure is still too painful. What he does do is to show that structures can't just be scaled up and still retain their original strength. He applies his arguments not just to ships and buildings but also to animals, pointing out

> the impossibility of increasing the size of structures to vast dimensions either in art or in nature; likewise the impossibility of building ships, palaces, or temples of enormous size in such a way that their oars, yards, beams, iron-bolts, and, in short, all their other parts will hold together; nor can nature produce trees of extraordinary size because the branches would break down under their own weight; so also it would be impossible to build up the bony structures of men, horses, or other animals so as to

hold together and perform their normal functions if these animals were to be increased enormously in height; for this increase in height can be accomplished only by employing a material which is harder and stronger than usual, or by enlarging the size of the bones, thus changing their shape until the form and appearance of the animals suggest a monstrosity.

In other words, as animals get bigger, their bones must get disproportionately thicker, as Galileo's own picture shows:

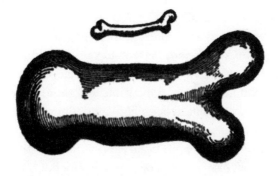

Figure 2.3. Galileo's diagram of a bone "whose natural length has been increased three times and whose thickness has been multiplied until, for a correspondingly large animal, it would perform the same function which the small bone performs for its small animal."

Galileo says, "From the figures here shown you can see how out of proportion the enlarged bone appears. Clearly, then, if one wishes to maintain in a great giant the same proportion of limb as that found in an ordinary man he must either find a harder and stronger material for making the bone or . . . admit a diminution in strength." Galileo's problem is reflected in the problems still experienced by modern-day surgeons of finding adequate materials for bone replacement in hips, knees, etc. SOURCE: University of Virginia Electronic Text Center, *Dialogue Concerning Two New Sciences,* trans. Henry Crew and Alfonso de Salvio (London: MacMillan & Co., 1914), available at University of Virginia Electronic Text Center, http://etext.lib.virginia.edu.

Galileo's scaling laws are still used by modern engineers, and his practical advice about building design is still useful for modern-day builders and do-it-yourselfers. He demonstrates, for example, that it is possible to have too many supports for a structure:

> SALVIATI: A large marble column was laid out so that its two ends rested each upon a piece of beam; a little later it occurred to a mechanic that, in order to be doubly sure of its not breaking in the middle by its own weight, it would be wise to lay a third support midway; this seemed to all an excellent idea; but the sequel showed that it was quite the opposite, for not many months passed before the column was found cracked and broken exactly above the new middle support.
>
> SIMPLICIO: A very remarkable and thoroughly unexpected accident, especially if caused by placing that new support in the middle.
>
> SALVIATI: Surely this is the explanation, and the moment the cause is known our surprise vanishes; for when the two pieces of the column were placed on level ground it was observed that one of the end beams had, after a long while, become decayed and sunken, but that the middle one remained hard and strong, thus causing one half of the column to project in the air without any support. Under these circumstances the body therefore behaved differently from what it would have done if supported only upon the first beams; because no matter how much they might have sunken the column would have gone with them.

We may no longer build our houses of marble, but Galileo's basic advice is just as relevant now as it was then. He even has something to offer when it comes to an object as simple as a piece of string. He showed that a piece of string, made of comparatively short fibers, remains strong because the twisting process used to make it presses the fibers together, increasing the friction between them so that they don't come apart when the string is pulled. How simple! How obvious! And how useful — if you are tying a plant to a garden stake with a piece of twine, for example, and the twine breaks, try twisting the twine to increase its strength before tying it. Don't twist it too much, though, or the very force of the twisting may break the fibers, as Galileo also pointed out.

Dialogues Concerning Two New Sciences was smuggled out of Italy and printed by the Dutch publisher Louis Elsevier, a publishing firm that still exists and has even published some of my own scientific papers. Elsevier was taking quite a risk, since he had sought advice from the Inquisition and had specifically been told that all of Galileo's writings were banned from publication, both in Italy and elsewhere. He was rewarded, however, by a bestseller that created a sensation in the European scientific community through its revelations of hitherto unknown laws and their practical applications. It contains the first description of the scaling laws that describe the strength of material structures, and which form the basis of modern architectural and engineering practice. Galileo also demonstrated in *Dialogues Concerning Two New Sciences* that objects moving at a constant velocity will keep on doing so even if nothing is pushing or pulling them, thus demolishing the old Aristotelian ideas at a stroke and providing a basis for Newton's First Law of Motion. This is not to say that everyone has caught on to the notion, even now. I described earlier how one group of experts had trouble with it when it came to moving a heavy magnet on wheels. Another "expert" who had trouble with the idea was Edward J. Churchill, the ballistics expert who testified at the famous Moat Farm murder trial of 1903 at Saffron Walden, near Cambridge in the U.K. Churchill stated categorically that the victim must have been shot in the head at close range, because the hole in the skull was fragmented, whereas, he said, the wound would have exhibited a clean round hole if the bullet had been fired from a distance, since the bullet would have had time to attain a greater speed after it left the barrel! Aristotle would have been proud to know Churchill, but it seems that Churchill had never heard of Galileo. Neither have modern cricket commentators who talk of a ball "speeding up" across the ground after it has left the bat. In fact, according to a recent survey, some 30 percent of people still have not grasped Galileo's principle.

I described in *How to Dunk a Doughnut* how I performed my own demonstration of the principle by riding my bicycle at a constant speed past my local pub and challenging onlookers to guess where a stone would land after I threw it straight up in the air. Some 30 percent guessed that it would land behind me, but Galileo's prediction that it would keep moving forward at the same speed even after I had let go of it was painfully confirmed when it landed on the top of my head. My experiment also confirmed Galileo's law of projectile motion, which was yet another revelation first published in *Two New Sciences,* and which is one of the most useful weapons in the modern scientist's armory. Most people know that when we throw a ball, or fire a shell from a cannon, the projectile follows a parabolic path. What few people seem to realize is that the path is made up of two components — the horizontal motion, where, according to Galileo, the projectile will keep moving forward at the same speed (so long as air resistance can be neglected) and the vertical motion, which is affected by gravity that eventually brings the projectile to earth. Galileo's law of projectile motion surprises many people by saying that the vertical and horizontal motions don't affect each other. So, for example, a bullet fired horizontally from a rifle will hit the ground at the same time as one that is simultaneously dropped from the hand.

Devotees of paintball games will find an immediate application for Galileo's law. Say that you are firing at an opponent who has just jumped from a tree. Should you aim at him as he jumps, or below him so that his path will intersect with that of the projectile? Many people would aim below, but the correct answer is to aim at the jumping point, because the paintball will drop at the same speed as the jumper.

Galileo's imprisonment might have seemed to be an inglorious and humiliating end to a glorious life, but without it, who knows what might have happened to Galileo and his work? It was while he was under house arrest that he finally wrote down the correct law of falling bodies, which he had studied experimentally in Padua for eighteen years, and which he added to the other insights that set science on its modern path. Would he have been able to escape the social whirl long enough to do this if he had not been imprisoned? Or would someone else have come along soon after and rediscovered those same insights? History suggests the latter, but we will never know. Galileo has been posthumously pardoned by the modern Church, but the Church of Galileo's time might have done all of us a favor by providing Galileo with the opportunity to transmit the insights that led to the new sciences that we still use today.

3

A Salute to Newton

Isaac Newton had a reputation for being right. In most cases, he was. We still use his three laws of motion, developed three hundred years ago, to predict the trajectories of rockets so accurately that we can land a man on the moon or an unmanned vehicle on the surface of Mars. NASA scientists used his law of gravity to swing the Galileo Space Probe once around Venus and twice around Earth to accelerate it for its final trip to Jupiter. We even use the calculus developed by Newton to perform the accurate calculations that are necessary for these and just about every other achievement of modern science.

There was just one subject in which Newton turned out to be wrong — the nature of light. He was convinced that it must be a stream of "fiery particles" that struck the eye to create the sensation of sight. The problem with this picture was that it couldn't convincingly explain what happens to light when it passes through a very narrow gap. You can see the for yourself by pressing the middle and index fingers of your right hand together, holding them up to a strong light in a sort of two-fingered salute, and then allowing the lower part of the fingers to separate slightly until the light just shines through. A series of dark lines will mysteriously appear in the gap, apparently suspended in space.

Newton thought that such dark lines arose when light particles were deflected by some unknown force as they passed close to an edge. It was only a suggestion, unsupported by experimental evidence, but Newton's iconic status was such that his picture went unchallenged for a hundred years. A London doctor named Thomas Young eventually did challenge it, suggesting that the light and dark regions could be accounted for much more easily if it was assumed that light travels like a wave in space, with light waves scattered from each finger edge interfering with each other as they cross (as do the ripples behind two swans passing on a lake), reinforcing each other in some places and canceling out in others. Young was violently attacked by Newton's supporters, and in particular by Henry Brougham Jr., an amateur scientist who later became famous for his outspoken opposition to the slave trade. Brougham was outraged that Young had dared to challenge his hero, and used his talent for journalistic invective to belittle Young and his work anonymously, with such effect that Young's burgeoning scientific career was brought to an abrupt halt. This chapter tells the story of the running battle between Brougham and Young and how Young's revolutionary "wave" idea was ignored by the British scientific establishment for nearly twenty years until it was vindicated in a totally unexpected way.

Thomas Young, born in 1773, was an unlikely candidate for the gossip and personal attacks that were to pursue him throughout his adult life. Even as a child he was terribly serious-minded. At the age of two he started to keep notes on every book he had read, and by the time he was four he had read the Bible from cover to cover twice. At the age of seven he was reading Newtonian philosophy, and by nine he had committed the *Westminster Greek Grammar* to memory and was reading Ewing's textbook on mathematics, "omitting gunnery" because of his Quaker upbringing.

It was not long before Young was reading and mastering the most important scientific book of the age — Newton's *Principia,* in which Newton announced his Law of Gravity and his Three Laws of Motion. Almost as an aside, Newton advanced an explanation for the curious behavior of the tides in the harbor at Hanoi, which was an important trading port for British ships in his day. His explanation proved to be a vital key to understanding the true nature of light, although it was Young, and not Newton, who eventually turned that key to unlock the secret.

Hanoi lies on the Red River, some fifty miles from the east coast of Vietnam and the Gulf of Tonkin. Newton's interest in its tides was stimulated after a seventeenth-century traveler named Francis Davenport sent a record of them to the Royal Society of London in 1678, asking the society to publish it as an aid to British trading vessels. Davenport explained that the harbor entrance was blocked by a sandbank, and only at the highest tides was the water deep enough for ships to cross. The tides followed an extraordinary fortnightly cycle. On some days there would be no tide, and the sandbank would remain uncovered. On the following days there would be just one flood tide per day, with each successive tide being higher until, after a week, the height was sufficient for ships to cross the sandbank safely. The height then dropped on subsequent days until, after a further week, there was again no tide at all.

Newton did not see Davenport's letter at first; it fell into the hands of Edmund Halley (discoverer of Halley's comet), who kept it to himself for six years while he tried to find an explanation for the tidal cycle. Eventually he devised a complicated equation relating the height of the tides to the position of the moon, which he published (along with Davenport's original letter) in the Royal Society's *Proceedings,* by which time numerous merchantmen had doubtless foundered on the Hanoi sandbank.

Halley's equation was what we would now call "empirical" — it described the behavior of the tides, but provided no reason for this behavior. He took his failure philosophically, saying that "to attempt to assign a reason [why the equation should be true] is a task too hard for my undertaking." Too hard for Halley, perhaps, but not for Newton, who immediately spotted that there must be two tides, coming from different directions. One, he suggested, came from the "Sea of China," with a delay of six hours. The other came from the "Indian sea," with a delay of twelve hours. If the two were of equal magnitude, and both high, they would add together to produce an enormous tide. If they were of equal magnitude, but one was high and the other was low, their heights would cancel and there would be no tide at all.

It was a clever piece of reasoning that he proudly published three years later in the *Principia.* With his masterpiece on mechanics produced, Newton then turned his attention to light, and within five years he had another masterpiece ready — *Opticks,* which he intended as his definitive work on the nature of light. Unfortunately, he made the mistake of leaving a candle burning in his room near the almost finished manuscript while he went to chapel at Trinity College, Cambridge. When he returned, his record of experiments and researches was reduced to a pile of ashes.

Newton set about redoing his experiments, and the restored version of *Opticks* finally appeared in 1704. In it, he described the original version of the "finger" experiment.

He placed two straight-edged knives together "so that their edges might be parallel, and look towards one another, and that the beam of Light might fall upon both of the Knives, and some part of it pass between their edges. And when the distance of their edges was about a 400th part of an inch, the stream parted in the middle, and left a Shadow between the two parts. This shadow was so black and dark that all the Light which passed between the Knives seem'd to be bent, and turn'd aside to the one hand or to the other."

Young would argue a hundred years later that the shadow arose because light was scattered from both knife edges, producing two sets of light waves that could reinforce or cancel each other at different places as they crossed, in the manner of the two tides that met at the Red River. Newton himself never saw the connection between the behavior of light and the behavior of the tides, because he simply could not believe that light came as waves. He had considered the possibility, but could not see how a wave picture could account for the ease with which light beams could be blocked off and then allowed to resume their passage, and argued that "because in the same place you may stop that which comes one moment, and let pass that which comes presently after [and] that part of Light which is stopp'd cannot be the same as that which is let pass."

Newton might not have believed in a wave picture, but as a good scientist he felt obliged to look for evidence in its favor. Unfortunately, his eyesight seems to have let him down. Newton had poor eyesight, and he failed to observe the narrow colored fringes that border the edges of a small hole through which white light is shone. If he had observed them, he would surely have gone on to make the vital breakthrough that only came a hundred years later, which was Young's recognition that "color" equates with "wavelength," as in figure 3.1.

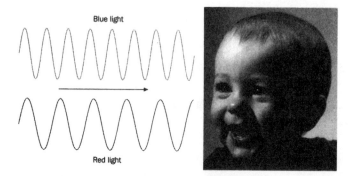

Figure 3.1. Different wavelengths are perceived as different colors.
The spectator is the author's grand-nephew Rowan, a budding scientist.

When light passes through a narrow gap, the edges of the gap act as secondary sources of light, and this light "interferes" with the light in the original beam, sometimes reinforcing its intensity and sometimes canceling it out entirely. If the light is all of one wavelength, an observer will see light and dark regions (interference fringes) in the light that has passed through the gap. If the light consists of a mixture of wavelengths (e.g., white light), these regions will occur in different places for different wavelengths, so that the white light appears as a series of closely spaced colored bands, visible to someone with good eyesight, though not visible to Newton.

When I was a child, trying to puzzle out for myself what light might be, I came to the same conclusion as Newton — that it must be a stream of particles hitting me in the eye. I tried to see them individually by shutting myself in a dark closet, hoping that they might be forced to squeeze through the crack at the edge of the door in single file so that I could experience them as individual flashes when they hit my eye. While I was in the closet I also tried shining two flashlight beams across each other, in the hope that some of the particles

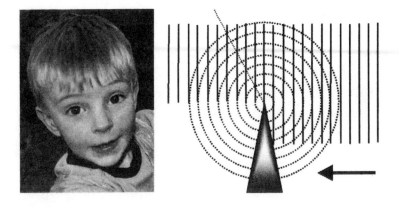

Figure 3.2. Interference fringes in light scattered from a sharp edge.
The parallel straight lines represent successive wave peaks of a "plane wave" traveling from right to left. When the light strikes the edge, the edge acts as a secondary source, sending circular ripples of light into space. These interfere constructively with the original wave only at certain angles (indicated by dotted straight line). The angle depends on the separation between the wave peaks (i.e., the wavelength), and so different colors are seen at different angles. The spectator is the author's grandson Gabriel, another budding scientist.

might collide and bounce around the closet, but my mother became suspicious of what I might be doing in the closet with two flashlights, and brought the experiment to a premature end.

I concluded that the particles must be extremely tiny, since so many of them could squeeze through the crack at once that I couldn't see them individually, and also because they didn't seem to hit each other when the two flashlight beams crossed. Newton, needless to say, went much further and concluded from his own experiments that light must consist of many different types of "fiery particle," each one corresponding to a different color, but all traveling at the same speed. He concluded from his observations that the particles would take around ten minutes to reach Earth from the Sun. It was an astonishingly close estimate; light actually takes just over eight minutes to reach us from the Sun. If Newton had been able to factor in the Earth–Sun distance, which was not known at the time, he would have been the first person to work out the speed of light, along with his other achievements.

Newton had a grave problem with his particle theory when it came to understanding the dark line between his knife blades, because there seemed to be no way that particles could suddenly disappear from the center of the beam as it passed through the gap. He could only guess that the trajectory of the particles was altered as they passed close to the knife blades. "Do not Bodies act upon Light at a distance," he asked, "and by their action bend its Rays; and is not this action *(caeteris paribus)* strongest at the least distance?" A similar conjecture would famously emerge from Einstein's General Theory of Relativity two centuries later, but the bending of light beams by gravitational forces was far too small an effect to account for the dark lines between Newton's knife blades.

Newton's conjecture was the first in a list of "queries" at the back of his book *Opticks*. These were questions that he had been unable to resolve for himself, but which he wrote down "in order to a farther search to be made by others." With time, though, his queries, like the rest of his writings, assumed the status of Holy Writ and were interpreted by many as statements that only a fool would challenge. One person who took them in this way was Brougham. Seventy years after Newton's death, Brougham set himself to confirm Newton's idea experimentally, and to show, "If a ray passes within a certain distance of any body, it is bent inwards [but that] If it passes at a still greater distance it is turned away." He didn't for a moment stop to consider that Newton might have been wrong.

When scientists set themselves to confirm a theory, rather than looking for evidence that the theory might be wrong, there can be only one outcome. So it was with Brougham, who triumphantly concluded that his results "proved" Newton's theory, when he had in fact introduced so many variable factors into his calculations that his results could have been fitted to any theory. He was a man with high social contacts, and he used one of them, Sir Charles Blagden, secretary of the Royal Society, to obtain publication of his results in the Society's *Proceedings* of 1796, with a further effusion in 1797.

Young had already clashed with Blagden when, at the age of twenty-one, he announced his discovery of the way in which the human eye can "accommodate" to focus on both near and far objects. He made his discovery shortly after he had begun to study medicine at St. Bartholomew's Hospital in London. Medical students of the time were given the eye of a recently slaughtered ox as a standard exercise in dissection. Thousands of students and their tutors must have done this dissection, but Young was the first to deduce that the muscular fibers surrounding the lens could increase its focal length by pulling on it to flatten it out. Always a believer in his own ideas, he submitted this one in an article to the Royal Society, whose members were so impressed they voted him a fellow the following year.

Blagden didn't believe that such a young man could have done such work, and spread the rumor that Young had plagiarized his idea from the great anatomist John Hunter. He claimed to have been told about the idea by Hunter, and to have then talked about it at a dinner party at which Young was present. Young reacted in a characteristic way by writing to every other guest at the party, asking whether they had any recollection of the discussion. One of those guests was James Boswell, whose compendious memory for conversational detail would surely have enabled him to recall the conversation. Another was Sir Joshua Reynolds, whose probity was unquestionable. Neither they nor any other guest could recollect any mention of the idea, and Blagden was forced to admit publicly that "he was by no means so clear as to be sure that he had told him Hunter's opinion."

Within three years the slander was forgotten, and Young was in high favor, albeit not for long. He had finished his medical studies in Göttingen and Edinburgh, and was now registered as a medical student at Cambridge so that he could be admitted as a licentiate by the London College of Physicians, a parochial body reluctant to recognize the "foreign" qualifications that Young had gained. A wealthy uncle had died, leaving him £10,000, his London house, and his medical practice. He had been invited by the Royal Society to deliver its premier lecture in the physical sciences, the Bakerian lecture "On the Mechanism of the Eye." He had even been appointed, while in his first year at Cambridge, as professor of natural philosophy at the Royal Institution in London, a newly formed institution whose worthy aim was to educate the masses in matters of science.

Young was appointed to the Royal Institution because of the incredible breadth of knowledge that had led his fellow students at Cambridge to call him "Phaenomenon Young." Young went about his task with characteristic earnestness. Within four months he had prepared and delivered some fifty lectures on subjects as diverse as "acoustics, optics, gravitation, astronomy, tides, the nature of heat, electricity, climate, animal life, vegetation, cohesion and capillary attraction of liquid, the hydrodynamics of reservoirs, canals and harbors, techniques of measurement, common forms of air and water pumps, and new ideas on energy." Unfortunately, he addressed his bemused audiences as though they were already deeply versed in the subjects, with the result that they didn't understand a word of what he was saying. Doubts had been voiced about the wisdom of educating the lower orders, but with Young at the helm there was not much danger of that.

Young's turgid lecture style suffered by comparison with the elegant presentations of people such as Humphrey Davy, and he lamented that "my audience has perhaps often been fatigued with insipidity or disgusted with inelegance." Even so, he went ahead to propose plans for a second series of lectures. The managers of the institution, however, were already turning it away from its original serious purpose into an intellectual playground for fashionable dilettantes and were only too aware of Young's unsuitability for such an undertaking. There was also the problem that Young was under attack by Brougham, who was angry that Young had cavalierly dismissed his experiments, but even angrier that Young had dismissed Newton in almost the same breath in his second and most important Bakerian lecture, "On the Theory of Light and Colours." Young had noticed that, when white light is reflected from a scratched surface, the scratches appear to be colored, and had said of these colors that "NEWTON has not noticed them. MAZEUS and Mr. BROUGHAM have made some experiments on the subject, yet without deriving any satisfactory conclusion."

Young believed that the colors could not be explained by Newton's particle theory, but only by his own wave theory. He provoked Brougham's ire still further by claiming Newton's explanation of the Red River tides as the stimulus for his theory, and quoting Newton himself as a supporter, saying, "A more extensive examination of NEWTON's various writings has shown me, that he was in reality the first that suggested such a theory as I shall endeavour to maintain [and] that his own opinions varied less from this theory than is now almost universally supposed."

We still use Young's picture of light as waves, with wavelengths extending from about 400 nanometers (violet) to 700 nanometers (red). Such short distances are hard to imagine, but other familiar forms of electromagnetic radiation have wavelengths that are more within our comprehension. Microwaves, for example, are similar to light waves except that they have wavelengths in the range of centimeters (i.e., around 100,000 times that of light), which is why they can't get through the millimeter-sized holes in the metallic coating on the door of your microwave oven. Radio waves have wavelengths of many meters, which is why they don't notice small objects, or even buildings, in their path (unless the objects are made of conducting material).

Young used data from Newton's own experiments to calculate the wavelengths of differently colored light, and showed how his theory could readily explain how white light reflected from a scratch became colored.

> Let there be in a given plane two reflecting points very near each other, and let the plane be so situated that the reflected image of a luminous object seen in it may appear to coincide with the points; then it is obvious that the length of the incident and reflected ray, taken together, is equal with respect to both points, considering them as capable of reflecting in all directions. Let one of the points be now depressed below the given plane;

then the whole path of the light reflected from it, will be lengthened. . . . If, therefore, equal undulations of given dimensions be reflected from two points, situated near enough to appear to the eye but as one, wherever this line is equal to half the breadth of a whole undulation, the reflection from the depressed point will so interfere with the reflection from the fixed point, that the progressive motion of the one will coincide with the retrograde motion of the other, and they will both be destroyed; but when this line is equal to the whole breadth of an undulation, the effect will be doubled.

The colors that Young described are called interference colors. We see them whenever a beam of white light (composed of all different colors) is reflected from two closely spaced surfaces, with the spacing between the surfaces being such that some colors (i.e., wavelengths) are reinforced, while others are canceled. The colors of a butterfly's wing, like the colors in light reflected from a scratch, arise from interference effects, as do the colors in an oil slick on the road or a soap bubble in the bath. Even more strikingly, the surface of a compact disc reflects different colors when it is tilted toward a source of white light.

Scientists such as myself routinely use interference colors to analyze the structures of small objects, but Young's explanation of their origin was too much for Brougham. Piqued by Young's dismissal of his work, and by what he saw as the misquoting of his hero Newton, he launched an anonymous personal attack on Young in the *Edinburgh Review,* an influential magazine he had cofounded with his friends Francis Jeffrey and the humorist Sydney Smith. The review would often use its influence in support of worthwhile causes, but on this

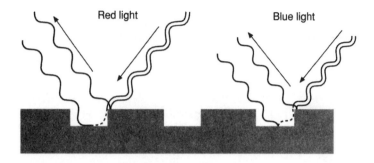

Figure 3.3. Origins of interference colors in white light reflected from the rectangular grooves in a compact disc.
Right, for the blue component of the light, the extra distance traveled by the light reflected from the bottom of the groove, compared to that reflected from the top is equal to the wavelength of the light, so the two reflected beams emerge "in phase" and add together to reinforce each other. *Left,* for the red component of the light, which has a longer wavelength than the blue component, the extra distance traveled is only half a wavelength, and so the two reflected beams are "out of phase" and cancel each other out. The overall effect is that the original white light appears blue after reflection. The extra distance that the light has to travel to reach the bottom of the groove depends on the angle at which the light strikes the disc, which is why different colors are seen as a disc is progressively tilted in white light.

occasion the influence was entirely malicious, especially when Brougham resurrected Blagden's long-dead rumor: "The author," Brougham said of Young, "introduced himself to the literary world, by a few desultory remarks upon a theory which he appeared to think new, but which had previously been exposed and refuted — the muscularity of the crystalline lens. Soon after this, he retracted his opinion [this was not true]; and a year or two ago he again brought it forward. . . . [W]e seriously recommend to him a due reflection upon the fact, in the history of his opinions. . . . Let it teach him a becoming caution in the *publication* of his theories."

Brougham used Young's own description of light as a vibration in a medium (which later scientists called the ether) to mock him: "[T]his paper contains nothing which deserves the name," he said,

> either of experiment or discovery, and . . . is destitute of every species of merit. . . . It is difficult to argue with an author whose mind is filled with a medium of so fickle and vibratory a nature. Were we to take the trouble to refute him, he might tell us, *My opinion is changed, and I have abandoned that Hypothesis: but here is another for you?* [Brougham's italics] We demand, if the world of science, which Newton once illuminated, is to be as changeable in its modes as the world of taste, which is directed by the nod of a silly woman, or a pampered fop? . . . [We recommend] to the Doctor . . . to invent various experiments upon this subject. As, however, the season is not favourable for optical observations, we recommend him to employ his winter months in reading the *Optics,* and some of the plainer parts of the *Principia.*

Phew! It is a mercy, one might think, that science is now a more dispassionate activity, and that such personal attacks are a thing of the past. If one thought this, one would be wrong. Science is still a very human activity, and human emotions are still very much to the fore. One of my most cherished memories is of a seminar given by a distinguished visitor to an Australian research institute. The head of the institute disagreed with his visitor's theories, and five minutes after the start of the seminar he stood up from his position in the front row, said "Bullshit!" in a loud voice, and walked out, leaving the poor visitor to flounder for the rest of his allotted hour.

At least the institute head concerned had the courage to voice his disagreement publicly. The anonymity of Brougham's attack made it very difficult for poor Young to reply. He did his best in a privately printed pamphlet where he argued his case at length. It was a technique that had worked for others, notably Lord Byron. When Byron's poetry was lambasted by an anonymous article in the *Review*, he responded in verse in a satirical pamphlet entitled *English Bards and Scotch Reviewers*, which ran swiftly through several editions and made its author famous. Young's turgid pamphlet met with a different fate. It sold precisely one copy.

Young's other response was to carry on and produce new evidence for his theory, in an experiment that anyone could try at home. In Young's own words: "I made a small hole in a window shutter, and covered it with a piece of thick paper, which I perforated with a fine needle. . . . I brought into the sunbeam [passing through the hole] a slip of card, about a thirtieth of an inch in breadth, and observed its shadow. Besides the fringes of colours on each side of the shadow, the shadow itself was divided by similar parallel fringes."

Explain that one, Newton! There is no way that the "particle" theory can account for this effect, but Young could, in the insight for which he is best remembered: "[N]ow these fringes were the joint effects of the portions of light passing on each side of the slip of card, and inflected, or rather diffracted, into the shadow."

Young had finally caught on to the point that he had missed in his earlier lectures, where he had talked about *different* sources of light interfering with each other. Now he saw that most interference effects arise when just *one* source of light becomes divided in two, for example by being reflected from the back and front surfaces of an oil film or a soap bubble, with the two beams later recombining after having traveled different distances.

His proof came in the form of an ingenious trick where he blocked the light beam passing on one or other side of the card, and found that the fringes promptly disappeared. In other words, light diffracted from *both sides of the card* was necessary for the fringes to form in the shadow. It was an amazing observation, and Young was careful to say, "I send you not this account barely upon the credit of my own eyes; *for there was a clergyman* [my italics] and four other gentlemen in company, whom I desired to view the colours attentively."

Young's brilliant experiment should have convinced anyone — even Brougham, if he had chosen to repeat it for himself. Brougham, however, simply refused to believe Young's results (just as Guilo Libri had refused to look through Galileo's telescope) and renewed his personal attack on Young. "The fact is," he said, "we believe the experiment was inaccurately made; and we have not the least doubt, that if carefully repeated, it will be found either that the rays, when inflected, cross each other and thus form fringes. . . . Or that, in stopping one portion, Dr. Young unwittingly stopped both portions; a thing extremely likely, where the hand had only one-thirtieth of an inch to move in."

Such accusations of experimental incompetence are not uncommon, even today, and are all too often used as a first line of defense when workers disagree. The disputes are usually resolved through further experiment, but authority and invective can hold sway for a surprising length of time. In Young's case, according to his biographer, "The effect which these powerful and repeated attacks produced upon the estimate of Dr. Young's scientific character was remarkable. The poison sank deep into the public mind, and found no antidote. . . . We consequently find that the subject of Dr. Young's researches remained absolutely unnoticed by men of science for many years."

Even when Young produced ultimate, dramatic confirmation of his wave theory, the British scientific establishment ignored it. The confirmation, still quoted in modern textbooks, was in the form of his famous "two-pinhole" experiment, which he published in 1807 in his *Lectures on Natural Philosophy*, six years after he had first proposed the idea of light as waves. All he did was to shine light on a screen that had two tiny pinholes close together and look to see what happened to the light on the other side. "When the two newly formed beams are received on a surface," he announced in triumph, "their light is divided by dark stripes."

Many people have repeated Young's experiment without knowing it when they have held an umbrella up to a streetlamp on a dark night. Every gap between the threads constitutes a tiny hole that acts as a fresh source of light, producing a wonderful array of interference patterns as the light beams recombine. They can only be accounted for on the assumption that light behaves as a wave that is scattered from each side of each thread. Young's experiment now appears in physics textbooks as the final nail in the coffin of the old-fashioned "particle theory" of light, but the truth is that Brougham's attacks had been so effective that the significance of Young's experiment was ignored for a further eleven years. It might have been ignored for even longer but for the unusual circumstance that Young had been a child prodigy who had mastered eight languages by the age of fourteen. It was because of this, and not because of his scientific prowess, that he had been elected as foreign secretary of the Royal Society at the age of twenty-one. His position put him into contact with many leading overseas scientists, and in a letter to one of them, the French physicist Dominique Arago, he compared the motion of light to that of the wave that passes down a stretched cord when the end is shaken up and down. It was the final clue.

Up until then everyone, including Young, had thought of the vibrations of light waves as being backward and forward in the same direction as the beam. This is the way that sound waves travel, as Young knew. Even with the example of the tides at Hanoi harbor before him, Young had still not taken their shape literally and had stuck with the sound analogy.

Sound waves are *longitudinal,* i.e., the vibration is in the direction of propagation. Water waves, and the waves in a stretched cord, are *transverse,* i.e., the wave motion is perpendicular to its direction of propagation. Australians such as myself have an unfair advantage when it comes to understanding this idea, since our time spent swimming among ocean waves has taught us that a nonbreaking wave will lift us up and down like a bobbing cork but will not carry us forward.

When Young's letter to Arago arrived in April of 1818, he showed it to a young compatriot, Augustin Fresnel, who immediately saw that it could explain the puzzling phenomenon of *polarization,* where a single light beam seems to be split in two when it passes through certain transparent materials, of which Iceland spar (fluorite) was then the best-known example. Newton tried to explain polarization away by assuming that his "fiery particles" had two sides with different properties. Young couldn't explain it at all. Now Fresnel saw that it could be explained if a light beam was composed of transverse waves, vibrating perpendicular to the direction of propagation.

The easiest way to understand this concept is to consider how Polaroid sunglasses work. The lenses are designed to allow light that is vertically polarized (i.e., vibrating up and down with respect to the ground) to pass, but to cut out that which is horizontally polarized (i.e., vibrating side to side with respect to the ground). Most light is a mixture of both, but light that is reflected from the ground or from the surface of water (i.e., glare) is largely horizontally polarized, which is why Polaroid sunglasses are so effective at cutting out glare — so long as the wearer is sitting or standing upright. Someone lying on his side and looking through Polaroid sunglasses will have rotated the lenses by 90 degrees, and will receive an eye full of glare.

Fresnel provided the basis for the invention of Polaroid sunglasses when he presented a mathematical theory of Young's "transverse wave" idea in a prize essay submitted to the Académie des Sciences. The academicians who examined the essay were all believers in Newton's particle theory, and they must have been delighted when one of their number, Siméon Poisson, pointed out that Fresnel's theory made the ridiculous-sounding prediction that the shadow cast by a circular object should have a bright spot right in the center. But Poisson was an objective judge, and he arranged with Fresnel to try the experiment. To his astonishment, the bright spot appeared, and Fresnel was awarded the prize.

We use the "bright spot" idea in the overhead projector lens (also known as the Fresnel lens), which consists of a cunningly arranged series of alternate transparent and opaque concentric rings on a flat bit of plastic, spaced according to Fresnel's theory so that multiple "bright spots" cooperate to form an image of whatever is placed on the plate. After the success of his essay, Fresnel continued to develop the wave theory of light for eight more years, bringing it to such a pitch of perfection that the particle theory was finally vanquished — for everyone except Brougham, that is. A quarter of a century later, he was still doing experiments designed to prove that Young was wrong. Not that he admitted it. In his last paper, published in 1850, he began by saying, "I have purposely avoided all arguments and suggestions upon the two rival theories — the Newtonian or Atomic, and the Undulatory," thus putting the two on the same footing, an old political trick. Then, in the very next paragraph, he assumed that light rays undergo "flexion" as they pass near bodies — either "*inflexion,* or the bending towards the body [or] *deflexion,* or the bending from the body." One has to admire his persistence. He even attempted to top Young's introduction of a clergyman as a witness by introducing "my friend Lord DOURO, who has, I believe, hereditarily, great acuteness of vision," as his own witness.

For everyone else, the fight was long over. The final act had come when Fresnel was awarded the prestigious Rumford Medal of the Royal Society and Young, back in favor once more, was given the task of forwarding the medal to him. Sadly, Fresnel was now desperately ill, and a week later he died. His name lives on in the Fresnel lens, while Young's name lives on in Young's modulus, a number that characterizes the elasticity of a material. My father once tried to surprise me by explaining that glass is one of the most elastic of materials — what he meant was that, if it is deformed slightly, it returns very closely to its original shape. I was to use that information many years later in my own research, when I made glass springs to measure very small forces. You can use a glass in another way, to discover one of the more surprising outcomes of Fresnel's mathematics, expanded and improved by James Clerk Maxwell. Fill a wineglass three-quarters full, hold it by the bowl about halfway down, and tilt it slowly while you are looking at an angle through the surface of the wine. Suddenly your fingerprint will come into view, but not the finger behind it, because light that comes from glass to air, and which is totally reflected at the glass-air boundary, penetrates a tiny way into the air before returning, illuminating only that part of the finger in close contact with the glass.

Now that you have the glass in your hand, you might also like to drink a toast to Thomas Young and Augustin Fresnel — and perhaps even to Henry Brougham as well.

Envoi: The Modern Picture of Light

The dispute between Brougham (representing Newton) and Young was a dispute between two metaphors about how information can be transmitted — either by sending a messenger (a particle) or by creating a disturbance in a medium (the ether) that fills the space between the sender and the receiver. It eventually transpired that there *are* individual messenger particles, though not of the sort envisaged by Newton. They are called photons, and my childhood experiment to see individual particles passing through a crack in a closet door was not as silly as it sounds. If we progressively reduce the intensity of a light source, and use a very sensitive detector called a photomultiplier, we eventually reach a stage where the light intensity becomes so low that we can detect and count individual photons as they strike the photomultiplier.

This does not mean that Young was wrong. When we try his "two-pinhole" experiment, with the light intensity made so low that only one photon at a time is passing through one or the other pinhole, we might expect that a photographic film placed behind the pair of pinholes would develop two spots, one corresponding to the photons that have passed through one pinhole and one corresponding to the photons that have passed through the other pinhole. Not so! What the film reveals is a series of interference fringes, just as Young predicted with his picture of light as waves.

This astounding result reveals (not for the first time) that Nature's actual behavior transcends the limits of our experience and consequently transcends the metaphors that we are able to derive from that experience. The mathematics of wave propagation still works, and we still use it. The picture of light as particles also works, and we still use that. Putting the two together is the province of quantum mechanics (see the appendix), and its interpretation has provided gainful employment for hundreds, if not thousands, of academic philosophers. It also provides a foundation for modern physics, but not one that can be grasped using common sense and metaphors. Working scientists do not attempt to understand it in that way; they simply accept that this is the way Nature works and use the rules that Nature imperiously imposes to understand and govern its behavior.

4

The Course of Lightning
through a Corset

People have always been fascinated by the power of lightning, and many ancient religions saw it as a tool of the gods. In Norse culture, lightning bolts were weapons of the god Thor, from whom we derive the word for Thursday, although there is no evidence that lightning is more frequent on Thursdays than on any other day. It was in fact a Sunday in 1895 when lightning struck the belfry of the Methodist Church in Quakertown, New Jersey, and traveled down the bell chain to wreak havoc among the congregation. It melted watch chains, burned skin, and cut every stitch of underclothing from the body of one Mrs. Burris without damaging any of her outer garments. It was especially drawn to Miss Minnie Frace, who was wearing steel hairpins and a steel-ribbed corset. It melted the hairpins, and traveled down the steel ribs of the corset to blow her out of her shoes, which are still preserved in the archives of the Greater New Jersey Annual Conference of the United Methodist Church.

If Minnie's corset had been placed on top of the steeple, and connected to the ground by a wire, its steel ribs might have protected the church and its congregation by drawing the lightning to them and passing it harmlessly to the earth. The ribs would have acted as lightning rods, which were proposed by Benjamin Franklin some hundred and forty years before the Quakertown event. Franklin favored pointed lightning rods, which he believed were best for attracting lightning. He was supported by the scientific establishment in America and England but was vigorously opposed by the Englishman Benjamin Wilson, who argued that lightning rods should have blunt, rounded ends. The controversy reached the highest levels of English society. Some of its members supported the scientific establishment, but others (including the king) supported Wilson. The scientists won the day, and only in the twenty-first century has it come to light that Wilson's seemingly perverse preference for blunted rods might have been right after all. This chapter tells the story of the acrimonious battle between the sharps and the blunts, and describes how modern understanding of lightning and electricity has finally led to a resolution of this age-old dispute.

What is lightning? The Greek playwright Aristophanes thought that it arose from a dry wind that "ascends to the Clouds and gets shut into them, . . . blows them out like a bladder [and] finally, being too confined, it bursts them, escapes with fierce violence and a roar to flash into flame by reason of its own impetuosity." The idea that lightning is heavenly fire wasn't just a product of his imagination; it had a solid logical basis. Lightning can set fire to trees, and even when it strikes a beach it can melt the sand to produce glassy tubes (up to 40 feet long) called fulgurites.

Aristophanes likened the roar of the accompanying thunder to a heavenly fart, a joke that got a good laugh from his audience. When I reinvented the joke as an eight-year-old, my father failed to see the humor, and seriously explained that the noise came from air rushing in to fill the vacuum left by the lightning stroke. He also told me that I could work out how far away a lightning flash was by counting the seconds while listening for the thunder, with every five seconds corresponding to a mile in distance. To this day I automatically start to count whenever I see a flash of lightning.

I nearly learned the hard way that the flash is actually a powerful electrical discharge. My father had built me a crystal set for my ninth birthday, and I can still remember the excitement with which I connected the aerial wire to the metal frame of my bed, ran an earth wire to the metal water pipe outside, and proceeded to apply a piece of sharpened wire evocatively called a "cat's whisker" to different spots on the germanium crystal while listening for a signal in the earphones as I snuggled under the bedclothes. When I struck a station that was broadcasting an England-Australia cricket test match, my excitement would have been complete except for annoying intermittent bursts of loud crackling in the earphones. My father explained that the noise came from lightning flashes in nearby storm clouds and made me take the earphones off until the storm had passed. He was wise to do so; some sixty people a year in Australia alone receive injuries when lightning strikes the line while they are using the telephone. My aerial wire, strung out to a nearby tree, was an equally vulnerable target. Shortly after I had taken the headphones off, a flash of lightning did hit a telegraph pole near our house, blacking out the neighborhood.

When Benjamin Franklin discovered the connection between electricity and lightning some 250 years ago, the only way to make electricity was to rub two different insulating materials together to produce "static electricity," much as a schoolchild might now electrify a plastic comb by rubbing it on the skirt or trousers. Franklin thought that the rubbing process transferred "electric fluid" from one material to the other. We now know that the "fluid" consists of individual electrons. When you rub a plastic comb on your clothing, you transfer several hundred electrons from the surface of the comb to the surface of the garment. A car driving along a road on a dry day does a more efficient job, and may rip thousands of electrons from the road surface. These distribute themselves over the car, and can give you a nasty shock as they try to find their way back to the road through your body when you reach for the door handle.

The rubbing process reaches its climax in clouds, where billions of water droplets are constantly striking against each other, with many losing or gaining electrons in the process. Those that have gained electrons find their way to the bottom of the cloud by a process we don't yet understand, while those that have lost electrons make their way to the top. The orphan electrons at the bottom of the cloud most often find their way back to the top by traveling together in a huge burst of electrical current, which so energizes the surrounding air molecules that they emit light to produce the characteristic blue flash. Such *intracloud* discharges account for more than 50 percent of all lightning flashes. Other flashes occur between one cloud and another, while a minority occur between the base of the cloud and the earth. It is these that most concern us because of the damage they cause when they start fires, strike buildings, or hit people or animals.

When lightning strikes a living body, most of it passes around the outside in a process called "external flashover." One curious example of this process was reported in 1776 when a pied (red and white) bullock was struck by lightning, which stripped off all of the white hairs and left the red hairs intact. It seems that white hairs are better conductors of electricity, which gives hope for people of my generation the next time we are caught in a lightning storm.

Another thing that provides hope is the history of U.S. park ranger Roy C. Sullivan, who survived a record seven lightning strikes between 1942 and 1983. The first lightning strike in 1942 happened as he was working up in a lookout tower and the lightning bolt shot through his leg and knocked his big toenail off. A second strike in 1969 burned off his eyebrows and knocked him unconscious, and a year later he was hit again as he was walking across his yard to get his mail. Things really picked up after that. In 1972 lightning set fire to his hair, and he had to throw a bucket of water over it. The hair had scarcely grown back when lightning set fire to it again, and three years later he was hit yet again. His final experience with lightning occurred in 1977, when he was hit while out fishing. None of these experiences with lightning appear to have dampened his spirit, which was only extinguished in 1983, when he was in his seventies and committed suicide after an unhappy love affair.

Roy Sullivan's body seems to have been peculiarly adapted to "external flashover." Those of us who are not so adapted could perhaps protect ourselves by an old alchemical recipe, which is to coat our bodies with the fat of a sea cow (known these days as a dugong), but dugongs are under threat, and in any case the risks are minimal. Only 20 percent of people who are hit by lightning die from the effects of the strike, usually because some of the current has passed through the heart, and relatively few people are directly killed each year by lightning: two thousand or so worldwide, about the same number as die from falling out of bed. More serious are the indirect effects, where lightning starts a fire or strikes a building. There was no effective way to protect against such catastrophes before Franklin's time. Churches were especially vulnerable because of their high steeples, and when Franklin came on the scene many American churches bore the additional danger of having gunpowder stored in their crypts. Measures to protect them from lightning during thunderstorms included violent ringing of the bells to try to break up the lightning strokes. Some medieval church bells still carry the inscription *fulgura frango,* meaning *I break up the lightning flashes,* but the most frequent result of ringing church bells during a lightning storm was the electrocution of the bell-ringer.

The problem was that no one knew what lightning actually *was.* The static electricity generated by rubbing two insulating materials together seemed in no way related to the enormous power of a lightning flash. Even the word *electricity* only means the ability to attract light bodies when excited by friction, and people before Franklin's time thought that this was the limit of its power. Franklin was able to recognize the potential power of electricity because of the invention of the Leyden jar by the Dutch physicist Pieter von Musschenbroek at the University of Leyden in 1746. The jar consisted of no more than two sheets of tinfoil covering the inner and outer surfaces of a glass vessel, but it could be used to store large quantities of electricity just as a cloud does, by carrying collections of opposite charges on the opposing faces. The charges could get back together if a conducting path was provided between the two faces, and a favorite party trick was to form the conducting path from a line of people who were holding hands. When the two end people respectively touched the outer surface of the jar and the tip of a metal rod connected to the inner surface of the jar, an electric shock would pass along the line and cause everyone to jump. King Louis XV of

France tried the experiment on a line of 180 of his knights, with gratifying results (from his point of view), and a similar experiment was performed on a line of Carthusian monks that was over a mile long.

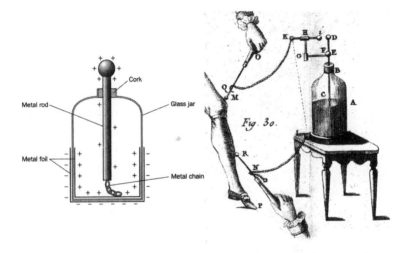

Figure 4.1. *Left,* **cross-section through a charged Leyden jar.** *Right,* **eighteenth-century French apparatus using a Leyden jar to apply an electric shock.**
When an electrical path is completed between the foil on the outside of the jar and the knob at the top of the jar, an electric current will flow.

Franklin thought of the electric current as a fluid that flowed along the conducting path. We now know that the "fluid" is composed of individual electrons that move in the opposite direction to that which Franklin guessed. Scientists and engineers still talk about an electric current "flowing" in the direction that Franklin proposed, but compensate for the wrong direction by saying that electrons have "negative" charge.

Franklin was fond of demonstrating the power of his Leyden jar by using a sharpened needle to draw large sparks from it, and he began to wonder whether lightning might be a similar spark on a much larger scale. To test his hypothesis, he proposed an experiment that seemed to smack of human sacrifice, where a man would be used to draw sparks from a cloud (!). The man was to stand in a "kind of sentry-box" on a high tower or a steeple, which had an iron rod "pass bending out of the door and then upright twenty or thirty feet, pointed very sharp at the end." The job of the man was to wait until a low cloud full of lightning came past, and then to hold his hand near the bottom of the iron rod in the hope that he could draw sparks from it.

Franklin records in his autobiography that his idea was "laughed at by the connoisseurs" in England, but it caused high excitement in France, and the famous comte de Buffon, keeper of the Jardin du Roi in Paris and one of the earliest evolutionary theorists, initiated a competition with his friends M. de Lor, "Master of Experimental Philosophy," and Thomas-François d'Alibard (who had translated Franklin's book into the French language), as to who would be the first to succeed with the experiment. D'Alibard won the race on May 10, 1752, with a very French modification (note the unusual use for wine bottles!) of Franklin's suggested experimental arrangement:

In a garden at Marly, six leagues from Paris, he set up an iron rod, an inch wide and forty feet long, pointed with brass. Having no cake of resin with which to insulate it from the ground, he used a stool which was merely a squared plank with three wine bottles for legs. At twenty minutes past two there was a single clap of thunder followed by hail. D'Alibard was just then absent. A former dragoon named Coiffer, left to watch the experiment, heard the thunder and hurried to the rod with an electric phial [a Leyden jar]. Sparks came from the iron with a crackling sound. Coiffer sent a child for the prior of Marly, who had heard the thunder and was already on his way. Meeting the child in the road, he began to run. The villagers, believing that Coiffer had been killed, ran after the prior through the beating hail. Terrified, they stood back ten or a dozen paces from the rod, but in broad daylight they could see the sparks and hear the crackling while Raulet the prior drew off all the electric fire. He sat down and wrote a letter which Coiffer took to d'Alibard, who three days later made his report to the Académie Royale des Sciences. Following the course which Franklin had outlined, he said, he had arrived at incontestable proof. Franklin's idea was no longer a conjecture.

By the time Franklin heard about d'Alibard's success, he had devised and tried a rather safer version of the experiment — his famous kite experiment, where he flew a silk kite in a storm cloud and drew sparks from a brass key attached to the bottom of the string. The kite string conducted electricity because it was wet, but Franklin himself stayed indoors, with the string passing out of an open window, so that the part inside remained dry and electrically insulating. The only electricity that could then reach him was in the form of sparks from the key attached to the wet part of the string.

People have been killed trying Franklin's experiment in the open, which is not a good place to be when lightning is about. I remember an occasion when I was visiting the Aspen Center for Physics, high up in the Rocky Mountains of Colorado, and had taken a walk in the mountains in the company of a well-known physicist. We were near the top, some 13,000 feet above sea level, when lightning began to strike *upward* from the clouds in the valley below. My companion was carrying a steel-shafted umbrella, and to this day I will swear that I could see an electrical corona forming around the tip of the shaft. When she turned to talk to me, I was a hundred yards away, attempting to communicate through shouting that she should drop the umbrella and run for it.

The corona was evidence that electrical charges can have substantial long-distance effects, even when they are not flowing in the form of a current along a wire or a spark through the air. Franklin found that a sharp-pointed needle could slowly draw the charge from a Leyden jar even in the absence of a spark, and believed that pointed metal rods could be used to discharge clouds in a similar way, emptying them of their electricity and thus obviating the possibility of a subsequent lightning strike. Under his enthusiastic promotion, pointed metal rods, connected electrically to the ground (i.e., "grounded"), were fitted to many public buildings in his hometown of Philadelphia. He fitted one to the roof of his own house but frightened the life out of his wife, Deborah, by arranging for the ground wire to pass through the *inside* of the house and past their bedroom door, where he had left a six-inch gap that sparks would jump across during an electrical storm, ringing a brass bell in the process. His wife was so frightened of the arrangement that she contacted him once when he was visiting England, asking in despair how to switch it off.

English society was slower to adopt Franklin's rods, but a committee composed of fellows of the Royal Society did cause one to be placed on top of the government gunpowder magazine at Purfleet-on-Thames near London in 1772. There was just one dissenting voice on the committee — that of

Benjamin Wilson, who argued that Franklin's idea that lightning rods work by quietly discharging a cloud was nonsense:

> If those gentlemen, who argued at the committee for the *necessity of points,* could have made it appear, that such points draw off and conduct away, the lightning *imperceptibly* and *by degrees, without causing any explosion,* during a thunderstorm (which seems once to have been the opinion of Dr. Franklin) I should readily have subscribed to their report.
>
> But experience shows us, that the fact is otherwise: there being many instances, where violent explosions of lightning have happened to conductors that were sharply pointed.

Wilson was dead right. Lightning rods do not slowly drain electricity from clouds to render them harmless, as Franklin and the rest of the scientific establishment believed. As Franklin later came to realize, lightning rods act by inducing bolts of lightning to form and strike the rod. Wilson recognized this and was alarmed at the idea of deliberately provoking lightning strikes above a building as vulnerable as a gunpowder magazine. He was even more scathing when lightning struck the Purfleet magazine in 1777 but missed the lightning rod entirely, striking instead an iron cramp some fifteen meters away and leaving the occupants "much frightened," but fortunately not blown to pieces. According to Edward Nickson, the Purfleet storekeeper, "If the conductor on the house has acted, it is imperceptible as I am informed."

The Royal Society committee was promptly reconvened, and came to the less-than-obvious conclusion that, since one Franklin rod had failed, the building and its neighbors should be fitted with fifty-one more. Wilson thought that his fellow committee members must have taken leave of their senses. He believed in lightning rods that would not provoke a strike (as he thought that sharp ones were bound to do), but which could carry anything that the clouds might throw at them if a strike *should* happen to occur. He thus thought that lightning rods should be blunt, and of sufficient thickness to carry large amounts of electricity. When his passionate minority report had no effect, he took a course that many a holder of a minority viewpoint has since taken — he went public.

Wilson's way of going public was to build a scale model of the Purfleet magazine and bombard it with artificial lightning. He staged his experiments in the ballroom of London's grand Pantheon, an elaborate rotunda in Oxford Street that was built along the lines of the Hagia Sophia in Constantinople. Such was the public interest that the proprietors of the Pantheon were able to charge the equivalent of £13 ($23) in today's money for people to come and watch. The public weren't the only ones to be interested — members of the Board of Ordnance and even the king himself watched many of the experiments, whose scale makes impressive reading even today. The "lightning" came from a long tube that was connected to a machine that could generate static electricity at the turn of a handle. The tube, made from wooden drums, was 47 meters long and 40 centimeters in diameter, and was covered with 37 kilograms of tinfoil. It was suspended from the ceiling by silk threads, and for some experiments a further 2 kilometers of wire was attached, strung around the Pantheon in great loops suspended from the ceiling by more silken threads. The size of Wilson's model building was puny by comparison, being only 60 centimeters wide by 45 centimeters deep by 40 centimeters high, and fitted with pieces of wire and scraps of tinfoil to represent downspouts, guttering, and lightning rods. The model was mounted on a carriage so that it could be wheeled under the suspended cylinder until it came close enough to draw a spark. Using it, Wilson proved to his own satisfaction that round-tipped rods fitted to the roof of the model had to come four to twenty times closer to the charged cylinder than did sharp-tipped rods before they would draw a spark, with no evidence that the cylinder was discharged in the absence of a spark. The other committee members were less convinced, partly because Wilson had no way to measure the electrical charge on his cylinder except by getting someone to touch it and tell him how strong the shock had been. This led to some problems with his later experiments when he was limited to low charges because he "could not prevail upon anyone present at the time to take a higher charge" (!).

The main problem, though, seems to have been that Wilson was not a professional scientist. He was a portrait painter, whose presence on the committee was due to his position as "Painter to the Board of Ordnance." Even though his fellowship of the Royal Society was a result of successful electrical experiments that had earned him the prestigious Copley Medal, he was still regarded as an amateur interfering in professional affairs. His position was further undermined when Jean Hyacinth de Magellan, a Portuguese scientific enthusiast then resident in London, wrote to the secretary to the St. Petersburg Academy of Sciences, which had recently bestowed a medal on Wilson, that Wilson's experiments were fraudulent, and that scientific observers believed that Wilson had been surreptitiously making and breaking electrical connections to induce the sparks. We will never know whether these accusations were true. Wilson's basic observations have been verified many times since, but he would not be the only scientist to have faked results he needn't have faked. In any case, the point is immaterial. Whether he faked some results or not, his scientific reputation was destroyed, and even though he continued to urge his point for some years, his chance of carrying the day with the scientific community had vanished.

Wilson had another string to his bow, however. As "Sergeant Painter to the King," he had the ear of King George III, who was incensed by Franklin's role in the American War of Independence, and who reportedly defied the scientific establishment by ordering that Franklin's sharp-pointed lightning rods should either be removed from royal buildings or fitted with cannonballs. The president of the Royal Society is said to have tartly commented: "His Majesty could change the laws of the land at will, but could not reverse or alter the laws of nature," and he and his fellow committee members continued with their recommendation to fit a plethora of electrically linked Franklin rods to the roof of the Purfleet magazine.

So Wilson lost the battle, but, as with many losers in scientific disputes, he had had a real point to make. It did not depend on his own experiments but only on experiments that Nature had made, and which Franklin himself had quoted:

> Buildings that have their roofs covered with lead or other metal, and spouts of metal continued from the roof into the ground to carry off the water, are never hurt by lightning; as, whenever it falls onto such a building, it passes in the metal, not the walls.

In other words, people already knew that a metal downspout full of water could carry a lightning strike safely to earth. All that Wilson had wanted to do was to fit buildings with blunt metal rods that were the electrical equivalent of downspouts. He correctly argued that no one knew how much "electric fluid" was contained in a cloud, and that there was an unquantified risk that if the lightning rods were pointed, they would draw to themselves more than the ordinary power of a lightning strike — perhaps much more. His experiments in the Pantheon did not prove his point, but by showing that pointed rods could draw sparks to them from a longer distance than blunt rods, the experiments at least gave his point some substance.

It was another sixty years before the beginnings of a true description of the difference between blunt and pointed rods began to emerge. The key to picturing the difference came from the brilliant Michael Faraday, who conceived the idea that all electrical charges produce an electric "field" through which they can exert a force on other charges. The simple conceptual way to test for the presence of such a field is to place a single positive charge in different positions and measure the force on it. If this test were applied to Wilson's experiments with sharp and blunt tips, the following picture would emerge:

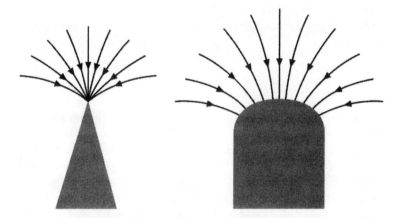

Figure 4.2. Schematic representation of electric fields near negatively charged sharp and blunt tips.

The direction of the arrows represents the direction of the force, and the density of the arrows represents the strength of the force. The lines become much more crowded near a pointed tip than they do near a rounded tip, showing that the forces are much stronger near the pointed tip. They are so strong, in fact, that they can tear molecules apart, splitting them into positively and negatively charged parts called ions. If the pointed tip is that of a lightning rod, the negatively charged lower surface of a nearby cloud will draw the positive ions to it. These collide with other atmospheric molecules to produce light and more ions, eventually resulting in a cascade of positive ions called a "leader" that initiates a lightning bolt by attracting a stream of negatively charged particles from the base of a cloud.

In the absence of a lightning rod, lightning bolts usually start from the base of a cloud and take the shortest path to the nearest available point of contact, which is why trees and church steeples are so often hit. The electrical conductivity of the object doesn't matter; a man is as likely to be hit as a bronze statue of equivalent size. The only thing that matters is height; if you are caught in a lightning storm, your best bet is to lie flat on the ground and put up with getting wet rather than hide under a tree or open your umbrella. Better still, get

Figure 4.3. The initial progress of a "normal" lightning strike.
A, time=zero, charge separation in cloud. *B, t=19 milliseconds,* branching, negatively charged "stepped leader" progresses toward ground. *C, t=20 ms,* upward-moving discharge begins from ground. *D, t=20.1 ms,* discharges meet ("attachment") some tens of meters above the ground and first return stroke is initiated. *E, t=20.2 ms,* return stroke, traveling at roughly one-third the speed of light, reaches cloud (another series of strokes follows some 40 ms later). Redrawn from Martin Uman, *The Lightning Discharge* (New York: Dover, 2001), 12.

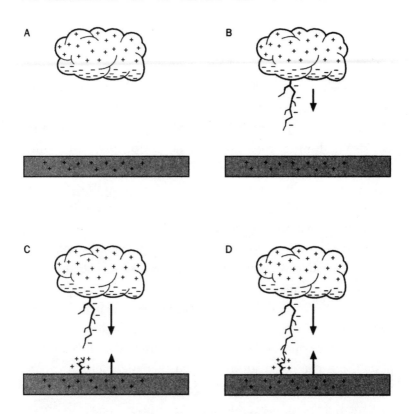

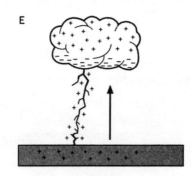

into your car but stay out of contact with the metal frame. Even if lightning hits the car, there will be no electric field *inside* the car (that a conducting shell completely excludes all electric fields was another of Faraday's remarkable discoveries).

Your best bet is to get the lightning to hit something else, and it would seem that Franklin was right in arguing that pointed lightning rods are best at performing this function. Modern science, though, has revealed a twist in the tale. When experimenters from the New Mexico Institute for Mining and Technology compared the lightning-attracting capacity of sharp-tipped rods with slightly blunted rods on the 11,000-foot South Baldy Peak in New Mexico, lightning struck the slightly blunted rods twelve times during one thunderstorm, but did not strike the sharp-tipped rods at all.

When the experimenters turned to calculation, they discovered the reason. The electric fields near the sharp tips were stronger, just as Franklin's and Wilson's results had suggested, but the calculations showed that the strength of the field near a sharp tip drops off very rapidly with distance, so that the field cannot produce enough ions in the air to create an effective "leader." The best rods seemed to involve a compromise, with the tip slightly blunted. The field at the tip surface becomes weaker, but this is more than compensated for by the fact that the field strength drops off much more slowly with distance, so that the field can affect a sufficient volume of air to form a proper leader.

One moral of this story is that intuition and common sense are seldom enough when it comes to scientific questions. Wilson's intuition turned out to be partly justified, but then so did Franklin's. There is a more serious moral to be drawn, however, and one highly relevant to many of today's scientific controversies. It concerns the assessment of risk. Wilson argued that no one knew how much electricity a cloud contained and that pointed rods risked bringing the lot down at once. His argument was strikingly similar to modern-day arguments about the risks posed by everything from nuclear power to genetically manipulated crops. There have even been claims that a vast new American particle accelerator on Long Island, the Relativistic Heavy Ion Collider (RHIC), could trigger the creation of a black hole that might swallow up our entire planet.

Scientists managing the new accelerator issued a rebuttal, saying that the risk of such a catastrophe was essentially zero. To my mind, this misses the point. No one could say that the risk was *exactly* zero, but no one could say that the risk of *not* building the instrument was exactly zero, either. What if the discoveries from it provided man's only means of salvation from some unknown future catastrophe?

Scientists who claim that risks are "negligible" are indulging in hubris, but so are politicians and the members of pressure groups who take it upon themselves to assert that risks are not worth taking. The truth is that people must take risks to progress, but the assessment of that risk and whether it is worth taking are often impossible to judge, as they were in the case of the lightning rod. Franklin's sharp rod may have protected the Purfleet magazine, or it may have led to its destruction, just as Minnie Frace's steel-boned corset may have drawn lightning to her but may have saved her life. We will never know. What we do know is that we must take risks to progress, but whether we take any particular risk is a matter for all of us and how we feel about the risk and its possible consequences. The role of science is to make sure that the feeling about both risk and the consequences is as factually informed as it can be. Science can do no more, but it should do no less.

5

Fool's Gold?

When I am introduced to strangers as a "chemist," most of them conclude either that I dispense prescriptions (as British chemists, also known as pharmacists, do) or that I spend my time in a smelly laboratory mixing chemicals together to see what will happen.

If I were an *al*chemist, they would be right in both cases. According to the *Oxford English Dictionary*, the word derives from the ancient Greek for an "infusion of plant juices," which is how the alchemists discovered and prepared many medicines, some of which we still use today. A strikingly similar word was used by the ancient Egyptians to describe a notorious Alexandrian sect that pursued chemical experimentation. The two words became one when they were taken over into the English language in the medieval word *alchemy*, which was concerned as much with the search for new drugs as it was with the pursuit of chemical transmutation. In both cases, however, the approach was the same — to mix, dissolve, and heat different combinations of materials in the hope that they would produce something novel or useful.

Some modern experts dispute the above etymology, but whatever the origins of the word, chemists such as myself are still successors to the ancient alchemists. We still search for ways to make new materials (including medicines) or to improve the properties of old ones by chemical or physical manipulation. The professed difference between our trade and that of the alchemists was spelled out by Robert Boyle in his 1661 book *The Sceptical Chymist*. Most alchemists were guided in their experiments by a belief that nature was resolvable into three principles — salt, sulphur, and mercury. Boyle claimed that this belief was unsupported by experimental evidence and argued that progress could only be made by setting such beliefs to one side. We should be skeptical, he said, of all statements concerning nature that are not directly based on experimental evidence, and we should pay particular heed to quantitative evidence, involving the accurate measurement of weight, volume, temperature, and other conditions.

This quantitative experimental approach has paid huge dividends. Boyle claimed that it put chemistry on a completely new footing and derided the alchemists for their previous haphazard efforts. The truth, recently uncovered by the American historian Lawrence Principe, was that alchemists were using quantitative experimentation hundreds of years before Boyle came on the scene, and that Boyle copied or appropriated from them many of the experiments and discoveries he claimed as his.

This is not to say that Boyle did not make many discoveries on his own account. He discovered new gases (including hydrogen and carbon dioxide) and a law is named after him that predicts the relationship between their volumes and pressures. He proposed the idea of blood transfusion, and discovered a chemical indicator (syrup of violets) that would indicate whether a material was an acid or an alkali by its change of color. Most important, from the point of view of his upperclass contemporaries, he discovered that ice could be made colder by mixing it with certain salts, and cardinals and others paid large amounts of money to have their drinks cooled by this method, which is still used in modern ice-cream makers.

Boyle wanted more, though. He wanted to be seen as "a virtuoso uninfluenced by older ideas," even when those ideas involved the very core of alchemy: transmutation. Principe has shown that Boyle, while ridiculing alchemy in public, secretly spent much of his life in private pursuit of the philosopher's stone, a mythical material that could supposedly "multiply" an initially small amount of gold indefinitely. This chapter traces his quest, with its curious mixture of modern chemistry and ancient alchemy, and reveals the debt that both Boyle and modern chemistry owe to the ancient art of alchemy.

The image of an alchemist in most people's minds is that shown in Joseph Wright's often-reproduced 1771 painting *The Alchymist in Search of the Philosophers' Stone Discovers Phosphorus*. The bearded alchemist (whose face bears a remarkable resemblance to that of Socrates in the Louvre portrait) is kneeling as if in worship, and gazing with Socratic intensity at a glass retort in which glowing white fumes are rising from a boiling liquid. The picture is imaginary, but the experiment and the alchemist were both real.

The alchemist was Henning Brandt, a failed Hamburg merchant who had turned to alchemy in an attempt to restore his fortunes. The experiment was his attempt in 1669 to manufacture the fabled philosopher's stone from sixty buckets of human urine. Needless to say, he failed, but by mixing the concentrated urine with charcoal and applying strong heat he succeeded in producing a new material — white phosphorus, one of the first elements to be isolated by chemical means.

Brandt's phosphorus glowed in the dark, an effect caused by its reaction with the small amount of oxygen left in the evacuated retort. When air was admitted, and the reaction speeded up, the phosphorus caught fire. Brandt found that it would not catch fire if it was stored underwater, which is the way that white phosphorus is still stored in laboratories today. The pure material was literally too hot to handle, but the gum that was left in the bottom of the retort contained only a small residue of phosphorus, and merely glowed when first exposed to air. It could be used to "write upon the Palm of your Hand, or upon Paper [and] what ever you write will appear all on fire, and the Letters may be read a long time after; but you must have a great care, that you do it softly, and to put it into Water, as soon as you have done, for if it happen to fire 'twill burn the place most dreadfully."

Brandt was keen to find a use for his miraculous new material. Its glow and the associated heat suggested medicinal properties, and it is possible that he saw it as a potential cure for syphilis. Whatever his reasons, they were sufficient for him to try an eye-watering experiment: "If the Privy Parts be therewith rubb'd," he said, "they will be inflamed and burning for a good while after."

Brandt's experiment seems foolhardy and ridiculous to modern eyes, but at least he tried it on himself and did not make exaggerated claims based solely on the glow. Some of the ancient alchemists, however, did make exaggerated claims, but most of them observed a due caution that a modern scientist would be proud of, and sought to verify their observations as carefully as they could. Their main problem was the lack of a good guiding theory (almost inevitable when scientists are breaking totally new ground), so that they didn't really understand what was going on. It was a problem that I, too, had when I first began to learn chemistry at school and came into contact with materials that would have delighted the ancient alchemists.

One of those materials was sulfuric acid, a powerful, corrosive solvent that was known to the ancients as oil of vitriol and was used by the alchemists in the "vitriolification" of metals and other materials. Schoolchildren these days would never be allowed access to such a dangerous substance, but we were each given a bottle of it in the wooden rack of chemicals provided for our experimental class. The rack also contained a bottle full of purple crystals of potassium permanganate. My mother would have recognized these as "Condy's crystals," which were kept in the family medicine cabinet in case of snakebite, but my mother was not there when I took a seat at the back of the chemistry class and began surreptitiously to drop the crystals into a beaker of concentrated sulfuric acid. Like the ancient alchemists, I wanted to see what would happen. Like them, unfortunately, I did not know enough about the underlying chemical mechanisms to predict what the outcome might be.

I was delighted when the crystals left twisted purple trails in the acid as they slowly sank. My neighbors were more interested in the acrid purple fumes that the mixture was emitting, and egged me on to add more crystals. As our ancient chemistry master, "Pop" Jane, rambled on at the front of the class, ignorant of what was going on at the back, I proceeded to pour the rest of the crystals into the acid, eventually producing nearly half a liter of hot, fuming purple liquid.

The alchemists used vitriolification to change the properties of a material added to the acid. I had unwittingly done the same, transforming the relatively harmless potassium permanganate into permanganic acid, a highly unstable liquid that can explode spontaneously. I found out later that the amount I had manufactured would have been enough to demolish the classroom. At the time, I was more concerned that the purple fumes would give me away and get me into trouble. I turned on the tap and cautiously flushed the acid down the sink, finishing just as Pop completed his spiel.

I had unknowingly taken a fearful risk because of my desire "just to see what would happen." The leading alchemists sometimes took equally dangerous risks, but their experiments were not nearly as purposeless as mine. They hoped, by using fire and corrosive solvents, to refine, break up, and recombine the component materials of the world so that they might produce other more useful or valuable materials. They were bold, and sometimes foolhardy, in the mixtures they used and in testing the properties of the many new materials they produced — but then, in the absence of a good guiding theory, they had to be. By taking such risks, they discovered the medicinal uses of many extracted plant materials, some of which we still use today. They purified antimony and other metals from their ores, manufactured calomel (sometimes still prescribed as a purgative when I was a child), sal volatile (smelling salts), and caustic soda, and synthesized many of the pigments used by artists. Above all, they developed or refined most of the chemical processes that we still use, including amalgamation, crystallization, filtration, precipitation, and distillation.

It is the last of these that is displayed in Wright's painting. A retort is a spherical glass vessel provided with a long exit tube. The vessel, containing the boiling urine, is hot, but the tube is well away from the source of the heat, and remains cool, so that the phosphorus vapor condenses on it (this is why the tube is called a condenser). Brandt never did succeed in finding a practical use for the phosphorus that he had produced, but he did manage to restore his fortune by selling the secret of its manufacture to the German physician Johannes Daniel Krafft, who showed off the new wonder substance around the courts of Europe. The manner of its production did not remain secret for long. Before long, the secret had "leaked out" (to use the infelicitous phrase of one commentator), and others began to copy and improve the process. One who did so was Robert Boyle, the independently wealthy son of the earl of Cork and founding member of the Royal Society. In 1680, eleven years after Brandt's discovery, Boyle wrote a sealed memo in which he described experiments in which he took "a *considerable quantity* of *Man's Urine*" of which "a good part at least, had been for a pretty while digested before it was used." It was then brought "to the consistence of a somewhat thick *Syrup*," mixed with sand, and distilled under gradually increasing heat to produce phosphorus. The parallels with Brandt's original process are striking, but nowhere in the memo (eventually published after his death in 1691) does Boyle acknowledge Brandt's priority.

The production of phosphorus required the application of a fire "as intense as the Furnace was capable of giving," a process that the alchemists called *calcination*. Fire was one of the strongest weapons in the armory of the alchemists, and they discovered that many materials could be broken down by a sufficient application of heat. They also discovered that the action of fire could sometimes be enhanced if the materials were placed on a block of charcoal and the fire blown onto them through a blowpipe.

The charcoal block and blowpipe were still in regular use as analytical tools when I began to study chemistry. A particularly effective demonstration of their power was with the red powder that the alchemists called *Mercurius praecipitatus,* which we know as mercuric oxide. When it was placed on a block of charcoal and heated with a blowpipe, the hot flame from the blowpipe drove the oxygen to combine with the charcoal (carbon) to form carbon dioxide, which disappeared into the atmosphere, leaving shiny globules of mercury behind. I would have been less impressed by this demonstration had I known just how toxic mercury fumes can be. The old alchemists were particularly vulnerable to its effects, since mercury was believed to be one of the *tria prima* of substances from which all materials were composed, and the preparation of a supposedly special form of mercury called "philosophical mercury" was seen as an essential first step in the preparation of the philosopher's stone.

It seems likely that the "severe nervous disorder" suffered by Newton in his later life was due to chronic mercury poisoning brought on by his alchemical experiments. In other cases, exposure to mercury fumes was deliberate. A common "cure" for syphilis in the seventeenth century consisted of placing the patient in a closed compartment, with only the head sticking out, and lighting a fire under a bowl of mercury over which the patient's private parts were dangling. No wonder Brandt was so keen to find something that could simply be rubbed on.

The closed compartment was called the *tub,* and its use was so common that Shakespeare was able to make oblique reference to it in *Measure for Measure* (act 3, scene 1) and to be confident that his audience knew what he was referring to:

> LUCIO: How doth my dear morsel thy mistress? Procures she still, ha?
> POMPEY: Troth, sir, she has eaten up all her beef, and she is herself in the tub.

The effectiveness of the tub was judged by the amount of saliva produced by the victim. Three *liters* a day was regarded as a satisfactory amount; any less, and the length and strength of the treatment would be increased.

Fire was not just used to heat mercury. It had many uses in alchemy, and almost all of them are still in use by chemists today. We use it, or its equivalent, in distillation, "the separation of a volatile component from a substance by heating so as to drive off the component as a vapor which is condensed and collected in a cooler part of the apparatus." We also use it in a variant of distillation invented by the alchemists and called "steam distillation" — a process in which plant materials are put in boiling water, with the steam carrying off the aromatic oils, which reappear as a golden yellow layer floating on top of the condensate produced by cooling the steam.

Steam distillation was one of the many chemical processes that I tried at home as a child. I put some eucalyptus leaves in boiling water in a saucepan on our electric stove and caught the steam on the lid so that water droplets would run down into a glass, with a layer of "eucalyptus oil" floating on the top. I was gratified by the production of the oil, but not nearly as gratified as I was by the result of another experiment that concerned water glass, a thick syrupy solution of sodium silicate whose main use was in preserving eggs. My science magic book, borrowed from the local library, suggested taking some unsuspecting person's box of matches, dipping the wooden stems in water glass, allowing them to dry, and then replacing them in the box. The unsuspecting person in this case was my father, a heavy smoker, but whose matches on this occasion simply refused to light, proving that sodium silicate is a very effective flame retardant, and also (from my father's reaction) that it is sometimes wise to keep one's mouth shut even when an experiment has turned out successfully.

Water glass holds a special place in the history of chemistry. It was discovered by the German alchemist Johann Rudolph Glauber (after whom Glauber's Salts are named). He manufactured it by fusing sand with sodium carbonate (i.e., washing soda) and then dissolving the glassy mass in water. A favorite (and, for once, safe) experiment among schoolchildren of my generation was to put crystals of materials such as blue copper sulphate or red cobalt nitrate in the bottom of a bottle filled with water glass. Over the course of several days, the crystals would produce fantastic growths that looked very much like colored plants or corals.

The person who first performed this remarkable experi-

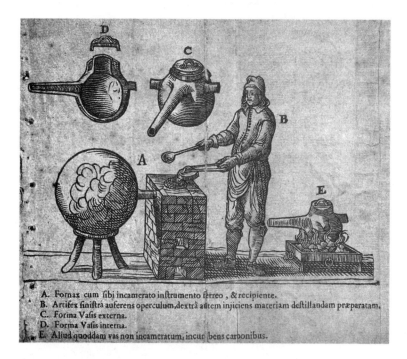

A. Fornax cum fibi incamerato inftrumento ferreo , & recipiente.
B. Artifex finiftrà auferens operculum,dextrâ autem injiciens materiam deftillandam præparatam.
C. Forma Vafis externa.
D. Forma Vafis interna.
E. Aliud quoddam vas non incameratum, incur bens carbonibus.

Figure 5.1. Furnace and distillation apparatus used by Glauber.
SOURCE: Johann Rudolf Glauber, *Works* (London, 1689).

ment was Glauber, and it was Glauber's description of the results that stimulated Robert Boyle's interest in chemistry. He
became fascinated by this "Liquor in which all Metalls grow
into Lovely Trees compos'd of Roote and Branches, & the
usuall Parts constituent of those Plants" and performed many
experiments with it. Boyle would later experiment with real
coral, which, he found, would produce gas bubbles when he
poured vinegar on it. The gas was carbon dioxide, one of
Boyle's truly original discoveries. It was produced from coral
because coral is mostly calcium carbonate, which releases carbon dioxide when it is exposed to an acid (in this case, the
acetic acid in the vinegar). Boyle found that the gas would not
support the life of some unlucky flies that he put into it. He
also discovered that it was heavier than air. My father once
showed me how a bottle full of carbon dioxide could be
"poured" over a candle to put it out (which it does for the
same reason as it kills flies, by denying them oxygen). I used to
delight in performing this simple trick as a child for the bene-
fit of my assembled aunts and uncles.

The "corals" produced by placing crystals in water glass
are not carbonates but silicates. Boyle later made out that he
had discovered them, and failed to reveal his alchemical source.
He also repeated Glauber's erroneous claim that alluvial gold
was produced in sand, again without acknowledgment. The
closest that Boyle came to admitting his debt to alchemy was in
his famous "Essay on Nitre," which he said was "an attempt to
make Chymical Experiments useful to Illustrate the Notions of
the Corpuscular Philosophy." In other words, according to Boyle,
the old alchemists might have discovered a few things by accident, but it took a "Philosopher" such as himself to devise and
interpret experiments that would reveal what was going on un-
derneath, and in particular to show that materials are made up

of very small particles that he called "corpuscles" and which we would now call atoms or molecules.

These acts of concealment were perpetuated by Boyle's followers after his death. His biographer, Thomas Birch, carefully omitted Boyle's alchemical correspondence from his list of collected works and may even have destroyed some of it to preserve Boyle's reputation as a "virtuoso uninfluenced by older ideas." Lawrence Principe has assembled an impressive amount of evidence which shows that both Boyle and modern chemistry owed a great deal to the alchemists. One thing they owe is the idea of weighing materials before and after they have been subjected to a process. The idea was developed into a rule by the alchemist Joan Baptista van Helmont, who believed that however much weight goes into a reaction at one end must come out at the other end — a belief that we now know as the Law of Conservation of Mass, one of the most fundamental and useful laws of modern chemistry. Van Helmont's most spectacular application of it was when he planted a willow tree weighing five pounds in a tub of earth that weighed two hundred pounds. After five years of watering with specially pure water, the tree weighed 163 pounds and 3 ounces, while the earth still weighed very close to two hundred pounds. Van Helmont concluded that the increase in weight was entirely due to water that had been transmuted into bark, leaves, etc. It was a reasonable conclusion, and only much later did scientists discover that much of the increase in weight came from the absorption of atmospheric carbon dioxide, transmuted by the plant (with the aid of sunlight) into organic molecules such as cellulose and chlorophyll.

Boyle's main debt was to one alchemist in particular — the Harvard-trained American George Starkey, whom Boyle met early in 1651, and whose recently discovered notebooks reveal just how close old alchemical practice was to modern chemical practice. Starkey pursued the doctrine that Boyle later claimed for his own, that "experimental results afford final judgment on the truth of conjectures"; in other words, fact is the death of hypothesis. Principe gives many examples of how Starkey and other alchemists designed and performed critical experiments to test their hypotheses, just as Boyle did and as modern scientists still do. Starkey's notes, unlike the directions presented in print by many of his alchemical contemporaries, describe his experiments so precisely that others can replicate them — another characteristic of modern science. One of the experiments he described was the formation of a "Philosophical Tree" from gold and mercury. The process began with the mixing of liquid mercury and amorphous gold in the bottom of a flask, and continued with heating the mixture in a specified way that caused the two substances to amalgamate and form a beautiful feathery tree that filled the flask. Principe has been able to replicate the process by following Starkey's directions, and there is no doubt that Boyle would have been able to do the same when Starkey sent a description of the experiment to him. Boyle's early fascination with "Metalls that grow into lovely Trees" was thus renewed, and the unusual effect of the mercury on the gold may have been one of the stimuli that accelerated his efforts to produce philosophical mercury.

"Philosophical mercury" was not what we now understand as mercury — that silvery liquid that my schoolfellows and I used to chase around the laboratory bench and try unsuccessfully to pick up with our fingers. Philosophical mercury was supposed to be an especially pure and very active form of mercury, contained in all materials and responsible for their properties. The philosophical mercury extracted from gold, for example, was responsible for its goldlike properties, and would convey those properties to other materials to which it was added. No wonder the alchemists were anxious to get their hands on it!

One of the properties that alchemical texts claimed for philosophical mercury was that it grew hot when amalgamated with gold. Boyle obtained a sample from "the only Operator I trusted in the making of it" (probably George Starkey), but only ten years after Starkey's death in 1665 did he eventually report that he had

> made trial of our Mercury, when I was all alone. For when no Body was by me, nor probably dreamt of what I was doing, I took to one part of the Mercury, sometimes half the weight and sometimes an equal weight of refin'd Gold reduced to a Calx or subtle Powder. This I put into the palm of my left hand, and putting the Mercury upon it, stirr'd it and press'd it a little with the finger of my right hand, by which the two Ingredients were easily mingled, and grew not only sensibly but considerably hot.

It is hard to know what to make of Boyle's description. Modern measurements show that the amalgamation of gold with mercury *absorbs* a small amount of heat (rather than emitting it), so that the amalgam should become slightly cooler as it is formed. This suggests that the mercury may have contained an impurity, which Newton suggested could have given the gold "a greater shock, & so put it into brisker motion." Modern chemistry (or, at least, the chemistry I was taught) provides no clue as to what such an impurity might have been. It is unlikely that Boyle was fooling himself, but he published no information on how the mercury might have been made. Professor Principe says that "the mystery of the mercury has sufficed to make it an object of curiosity for three centuries" and has informed me that he is using clues from Boyle's private correspondence in an attempt to solve the mystery, with promising results.

Mystery or no, the "incalescence" that Boyle experienced when he performed the amalgamation experiment left him in no doubt that he had the key to the philosopher's stone in his hand. Unfortunately, he didn't know how to use it. This was why he was forced to breach the secrecy that usually enshrouded his alchemical activities and to publish the result of his experiment. This action alarmed Sir Isaac Newton sufficiently for him to write to Boyle urging him to "preserve high silence" on this sensitive topic. But Boyle had no choice. Publication was his only way to attract the attention of the clandestine society of adepts whom Boyle believed to hold the secret of transmutation.

Publication, though, had its dangers, not the least of which was that Boyle might become identified as an alchemist himself. His answer was to publish his observations under a pseudonym, but the pseudonym he chose was unbelievably transparent. It simply consisted of his initials in reverse order, and it was as "B. R." that Boyle published his observations in the *Philosophical Transactions of the Royal Society.* He also took the unusual step of presenting his paper in two columns of parallel text, one in English and the other in Latin, presumably on the assumption that all alchemists would be able to read Latin, even if their native language was not English. To make doubly sure, he also published his article in German, the native language of many of the alchemists.

It seems extraordinary that the man who set out in public to demolish alchemy was so bent on pursuing it in private. A part of the answer was that Boyle didn't really set out to demolish alchemy, although later historians interpreted *The*

Sceptical Chymist as the book that established a clear and immediate distinction between ancient "unscientific" alchemy and modern scientific chemistry. Boyle was not challenging the alchemists' observations and measurements, however, only their interpretation in terms of the *tria prima*. His interpretation of his own and the alchemists' experiments was that materials are made up of elements, although his definition of an element was very far from the modern one. As he wrote in *The Sceptical Chymist,* "I now mean by elements . . . certain primitive and simple, or perfectly unmingled bodies; which not being made of any other bodies, or of one another, are the ingredients of which all those called perfectly mixed bodies [chemical compounds] are immediately compounded, and into which they are ultimately resolved." In other words, he believed that his "elements" were all aggregates of the same sorts of corpuscle, arranged in different ways, so that there seemed to be no barrier to the transmutation of one element into another by rearranging the corpuscles. The modern definition of an element also sees them as being constructed from "corpuscles" (at a simple level, electrons, protons, and neutrons), but these are arranged with the protons and neutrons in a tiny core, surrounded by a cloud of electrons whose number (which is exactly equal to the number of protons) defines which element it is. There is no possibility of transmutation unless the nucleus is disrupted by forces beyond the reach of chemists, whose effects are achieved only by manipulating the outermost of the electrons.

Boyle never did identify any particular material as an element. He even failed to identify hydrogen, the simplest of materials, as an element, although he was the first to discover this inflammable gas when he observed it bubbling up from the surface of some iron nails that he had put into dilute sulfuric acid. I improved on his experiment when I was at school and listening to another of Pop Jane's interminable rambles. I put some iron filings into dilute acid and added detergent to the mixture to capture the bubbles as they rose to the surface. Where I went wrong was in blowing the raft of bubbles through the air and toward a Bunsen burner, on the assumption that each individual bubble would explode with its own gentle pop. I should have foreseen that the whole lot would go off at once with a devastating bang, an event that brought the senior master running down the corridor, and which may have been one cause of Pop's premature retirement.

Another cause for Pop's retirement may have been the gradual diminution of the school's stock of chemicals after a friend and I volunteered to help in the storeroom. Our reward to ourselves was a tithe of every chemical, which we split fifty-fifty. I was very envious when my friend was able to augment his stock with a one-pound jar of arsenic trioxide that his grandfather had kept to kill ants. We used some of it to develop our own chemical test for arsenic, on the assumption that we would both grow up to be chemical detectives. We also laid our hands on a small bottle of mercury and tested its amalgamating properties on lumps of zinc, of which we had a plentiful supply.

We knew enough about the toxic properties of mercury vapor to perform our experiments in the open air, with the wind blowing away from us. Boyle recommended a similar approach to Starkey when the alchemist began to complain of "very horrid and seemingly Pestilential Symptomes" following experiments with mercury, antimony, and arsenic. Boyle's suggestion was that Starkey should remove the glass from the windows in his laboratory. He would have been better advised to suggest that Starkey should be more careful, since the alchemist often had accidents where glasses and crucibles were spilled or cracked, with the contents covering the walls and floor of his furnace. Starkey's financial position was such that he was forced to try to recover the material, which on one occasion involved disassembling the entire brick furnace.

In the main, though, Starkey was a careful experimenter. One of Boyle's main reasons for pursuing the elusive philosopher's stone was a painstaking demonstration in front of some of his friends in which Starkey succeeded in producing from antimony ore a material that looked very much like gold. It may even have been gold, since antimony ore often contains trace amounts of that element. Boyle also believed that he had himself achieved a form of reverse transmutation in his own laboratory, turning gold into silver by dissolving it in a highly corrosive mixture of nitric acid and antimony trichloride, and then fusing the resultant white powder with borax to produce silvery metallic globules. He called the acidic mixture his *menstruum peracutum,* or "universal solvent," although a translation as "ultra-effective solvent" might be more accurate; if it were a truly universal solvent, it is hard to see what he could have kept it in. The metal that he produced was probably pure antimony, or an alloy of antimony and gold. One thing that we can be sure of is that it wasn't silver, a pure element in its own right.

Boyle was persuaded by these experiments that he was on the track of the philosopher's stone, but the main reason for his pursuit of this goal seems to have been that he had seen a demonstration of the stone in action. According to his description, a traveling adept filled a crucible with lead in his presence, heated the crucible in a furnace until the lead had melted, and then cast some red powder on the surface of the molten lead. The adept then covered the crucible and replaced it in the furnace for fifteen minutes. When it was removed and allowed to cool, the lead had gone, to be replaced by a mass of pure gold. From that day on, Boyle *knew* (or believed that he knew) that transmutation was possible, and set himself to discover the secret.

The event probably occurred in 1679, some four years after the publication of Boyle's "philosophical mercury" paper, which seems to have achieved its intended purpose in attracting the attention of the charlatan. It seems hard to believe now that a scientist of Boyle's stature could have been so easily deceived by what must have been a trick, but scientists are as gullible as anyone else when it comes to being persuaded by evidence of something they want to believe in. Boyle never did realize he had been fooled, and he never did discover what the itinerant alchemist's secret was. Possibly it was a material that combined with silver, mercury, or lead to produce something that looked like gold. Many of the alchemists' preparations used sulphur at some stage, and many sulphides have a golden appearance. One such is iron pyrite (iron sulphide), which is known as fool's gold because of its ability to fool the uninitiated eye. Chalcopyrite (copper iron sulphide) is also a mineral that has a brassy luster, while antimony sulphide is the principal component of a golden pigment used by painters. There was plenty of scope, therefore, for the eye to be fooled by purely chemical changes in the alchemist's mixture.

By 1689, Boyle must have believed himself to be close to discovering the secret, because he sought and obtained the repeal of the three-hundred-year-old Act Against Multipliers, which outlawed transmutation "on pain of felony." He did actually succeed in producing a red powder that he believed had some of the requisite properties, and when he died in 1691, an anxious Sir Isaac Newton wrote to his executor, asking if the powder had been discovered among Boyle's effects. The executor, John Locke, unearthed the powder and sent it to Newton, but without any instructions for its use. Newton wrote again, asking for the "receipt," which Locke also managed to find. It was heavily coded, but Newton seems to have decoded it and tried it out. His attempt failed, which is hardly surprising, and Newton was thereafter dismissive of the whole idea of transmutation.

Boyle's recipes for transmutation were bound to fail, because the energies that he was using (those of fire or chemical change) were only powerful enough to affect the electrons at the surface of the atom. To achieve true transmutation, higher energies are needed that are sufficient to produce changes in the atom's nucleus. To date, the only way to effect such changes has been to bombard the nucleus with energetic particles produced by nuclear reactions or by giant particle accelerators.

Scientists have now found that a sufficiently powerful laser can also be used to do the job. Professor Ken Ledingham at the University of Strathclyde has used this method to turn gold atoms into mercury atoms — a joke on the old alchemists, but one with a serious purpose, since he has now adapted the

Figure 5.2. Interior of the large electron-positron collider (LEP) at CERN (European Organization for Nuclear Research) near Geneva.
The men are standing in the collider's doughnut-shaped tunnel (some 27 kilometers in circumference), where electrons and their anti-particles, positrons, were accelerated in opposite directions and allowed to collide head-on, annihilating each other and releasing a concentration of energy in the tiny collision volume comparable to that which existed in the universe a fraction of a second after its creation in the Big Bang. The machine has now been replaced by the even more powerful Large Hadron Collider, where protons are accelerated by powerful magnets to near the speed of light to crash into and disrupt atomic nuclei. SOURCE: CERN.

process to make radioisotopes for use in medical diagnosis (a similar process could also be used in principle to reduce the half-life of long-lived radioactive waste).

One of the goals of the ancient alchemists, that of transmutation, has thus been realized in a way that they could not possibly have foreseen. Their other goal — that of producing new materials by manipulating old ones — has been fulfilled many times by modern chemists, who owe a debt to Robert Boyle for emphasizing and codifying the need for reproducible, quantitative experimentation. It is a debt almost as great as that which Boyle and his modern successors owe to the ancient alchemists.

6

Frankenstein Lives

Movie audiences of the early 1930s gasped when Henry Frankenstein, played by Colin Clive, used a high-voltage surge of electricity to bring Boris Karloff to life as his all-too-realistic monster. Few would have realized that the scene was based on a real event, played out 130 years earlier before a public audience at London's Royal College of Surgeons.

The occasion was a demonstration of "galvanism" by the Italian scientist Giovanni Aldini. Aldini was the nephew of the physiologist and anatomist Luigi Galvani, and had acted as his uncle's assistant some twenty years earlier in a series of famous experiments where a dissected frog's leg was made to kick by the application of electricity. The effect became known as "galvanism," and after his uncle's death in 1798 Aldini traveled the world, putting on shows to demonstrate the effects of galvanism on everything from a severed ox head to a human body.

Aldini's London show was one of his most spectacular. When he applied a pair of electrodes to the body of the convicted murderer George Forster, which had been taken down from the gallows only an hour before, the legs kicked, one eye opened, and the murderer's clenched fist raised itself threateningly in the air. Members of the audience gasped as the body seemed to come back to life, and one lady fainted. Later demonstrations in imitation of Aldini's were even more impressive. The audience at an experiment in Glasgow scattered wildly when the application of an electric current caused the body's index finger to straighten out and appear to point to them one after another.

Aldini's demonstration was one of the inspirations for Mary Shelley's book *Frankenstein*. "Perhaps a corpse would be re-animated," she said in the preface; "galvanism had given token of such things." We still talk of people being "galvanized into action," which Galvani had certainly done when he applied the electric current from a spark generator to a dead frog's legs. But Galvani had gone far beyond that. He had found that he could induce the legs to kick without an electric generator, simply by touching the nerve and the muscle simultaneously with a pair of linked wires. This experiment convinced Galvani that animals must make their own electricity, but the physicists of his day (especially Alessandro Volta, after whom the volt is now named) were outraged by his conclusion. This chapter tells the story of the battle between Volta and Galvani and of its literally shocking outcome for the biological and physical sciences alike.

The effects of electricity on the human body were an endless source of fascination and speculation in the eighteenth and nineteenth centuries. The invention of the Leyden jar in 1745 (see chapter 3) had provided a ready, portable, powerful source of electricity, which was quickly applied to a rash of party tricks. One of these was a "magical picture," which was a "large metzotinto [*sic*] with a frame and glass, suppose of the KING (God preserve him)." The king's golden crown was hooked up to a hidden Leyden jar, and anyone who tried to remove it received a "terrible blow" through a series of gilt contacts.

A popular trick was to hook up a chain of people across a Leyden jar and watch them jump. It was while doing this that the Frenchman Joseph-Aignan Sigaud de la Fond made a remarkable discovery. In his version of the experiment, the first person in the chain touched the inside of the jar, and the last person brought his finger toward the outside, drawing a spark and causing everyone to jump as the electric current passed along the line. On this occasion, however, only the six people nearest to the person drawing the spark jumped. The sixth, a young man of delicate features, had failed to pass the current to his neighbor.

The rumor quickly spread through Paris that the young man was unable to transmit the current because he was not endowed "with everything that constitutes the distinctive character of a man." Sigaud subsequently repeated his experiment with three *castrati* in the line, all of whom jumped, but this was insufficient to quell the rumor, whose proponents took the view that there must be a difference in conducting power between "men who have been mutilated by Art and men towards whom Nature has displayed cruelty."

Sigaud performed many further "chain" experiments without obtaining a repeat of the effect. It looked as if the young man was in for a rough ride until, serendipitously, the effect recurred when Sigaud was experimenting with a line of sixteen people. The first few people jumped, but the last jumper again failed to pass the current along the line. Sigaud had the insight to look, not at the unfortunate man's testicles, but at his feet, and noticed that the man was standing on a patch of damp ground. He came to the brilliant conclusion that wet ground was a better conductor of electricity than the human body, and that the current was flowing to the ground instead of along the line. The young man's reputation was saved, and the ground was laid for the invention of the ground wire.

The idea of a ground wire is to provide an alternative, and easier, passage for an electric current to the ground rather than having it pass through your body. If it does pass through your body, the results can be spectacular, as I discovered when I was a visiting scientist working at the Physiological Laboratory at Cambridge University, where a disgruntled student had wired a Tesla coil (which produces around 40,000 volts) to the copper urinal habitually used by the professorial object of his disgruntlement. With the wet floor of the urinal, his body (and the accompanying liquid stream) formed an excellent conducting path from the urinal to the ground. According to one subsequent report, the professor managed not only to hit the tiles above the urinal, but also to wash the dust off a window some six feet higher up.

The observation that electricity could make people jump was not only used for party tricks; it became the focus for a series of increasingly bizarre attempts to find medical uses for electricity. One of the earliest, and most sensible, was to use it to cure paralysis. "If electricity can make active limbs move," asked its practitioners, "why should it not do the same for paralyzed limbs?" Exaggerated claims became the order of the day, and in the early 1750s one doctor in Montpellier was attracting patients at the rate of twenty a day to undergo electrical stimulation of their paralyzed limbs. Unfortunately for the patients, the miracle cure turned out to be no cure at all. Electricity could make their paralyzed muscles contract, but as soon as the source of electricity was removed the limb returned to its original state.

Other medical practitioners, both trained and untrained, claimed success in using electricity to "cure" just about every ailment under the sun. Gout, rheumatism, chilblains, diarrhea, deafness, and venereal disease were among the favorites, and electricity was even touted in America as a cure for piles, which must have been singularly uncomfortable for the patients involved. The prize for the most outrageous application of electricity to the human body, though, must surely go to the U.K., and to "Doctor" James Graham and his "Temple of Health."

Graham was an extraordinary figure. Born in Edinburgh in 1745, the son of a saddler, he claimed to have qualified as a physician at Edinburgh University, and even managed to practice medicine there for a time, although his claim to medical qualification was doubtful. From there he moved to America, where he seems to have learned enough about the purported effects of electricity to be emboldened to return to London and, in August 1779, open his Temple of Health in Adelphi Terrace. Here he strolled around, dressed in a white linen suit and carrying a posy of flowers, while he lectured his customers on the virtues of electricity, especially when it came to matters of sex. One of his assistants was the teenage Amy (Emma) Lyon (later to become the mistress of Horatio Nelson), who stood semi-naked among the statues as the goddess Hygiea, one of the "Goddesses of Youth and Health" who were there to encourage the customers' interest. If her poses did not work, she was wont to take a naked mud bath to further stir things along.

Graham had many electrical devices available to enhance his customers' sex lives. They could get a quick charge from the electrified "celestial throne," or bathe in water through which a charge had been passed. They could take a sniff of "electrical aether" (actually an extract of assorted plants that had supposedly been exposed to an electrical field) or drink some "aetherial basalm" ("electrical aether" with wine added). If these weren't enough, there was always the "celestial bed" — a huge contraption that could be tilted at different angles, and at whose head the words BE FRUITFUL, MULTIPLY AND REPLENISH THE EARTH sparkled with electrical fire. The injunction was intended for the users, childless couples who would spend up to five hundred pounds to make love in the bed and have their union blessed. To assist in the process, the mattress was stuffed with perfumed flowers and "sweet new wheat or oat straw." For wealthier customers, hair from the tails of English stallions was added to the mix. Most important for the users, however, was the purported effect of "15 cwt. of compound magnets . . . continually pouring forth in an ever flowing circle."

Graham may have been a charlatan, but the popularity of his establishment indicated the intense public fascination with the effects of electricity and magnetism on the living body. This interest was at its height in 1780 when the Italian anatomist Luigi Galvani, a professor at the University of Bologna, decided to test whether electricity might also have an effect on the limbs of dead animals. The animal he chose was the frog, which he dissected to remove the legs and attached spinal cord (this contained the nerves that stimulated the movement of the leg muscles). He then passed a small iron hook through the top of the spinal cord so that it could be hung from a pin or anchored onto a horizontal dissection board. The presence of this iron hook was to prove crucial in his coming dispute with the renowned physicist Alessandro Volta, and also in the modern interpretation of his experimental results, but its initial purpose was as a physical attachment point, and also as a place to which Galvani could conveniently attach a wire connected to a source of electricity.

To provide the electricity, Galvani mostly used a static

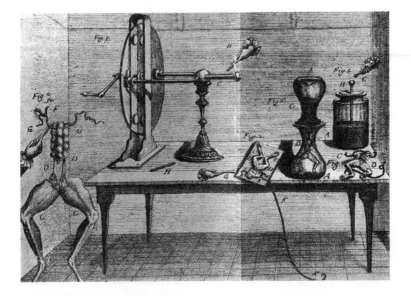

Figure 6.1. Plate 1 of Galvani's *Commentarius,* showing dissected frog's legs and spine with iron hook attached, as well as bits of electrical equipment from Galvani's laboratory.

electricity generator, with which the electricity was produced by continuously rubbing one material against another — a glass wheel rubbing against a leather belt, for example. As we now know, this process physically rips a small number of electrons from the surface of the leather and deposits them on the surface of the glass. The electrons repel each other, and spread out over the surface of the glass so as to get as far apart as possible. Such generators can still be found in modern science museums. Touching them literally makes your hair stand on end, because the electrons distribute themselves over your body, with those that reach the hair producing a sufficient repulsive force to separate the individual hairs.

When Galvani touched the generator with a wire running from the hook, he was disappointed at first to find that nothing happened. The legs were much heavier than individual human hairs, and the few electrons that found their way to the surface of the legs did not repel one another with sufficient force to push the legs apart. His disappointment turned to excitement when he withdrew the wire from the generator, producing a spark at the end of the wire near the machine and a substantial contraction of the frog's leg muscles at the other. A spark is a flow of electricity across a gap, and it seemed that static electricity alone was insufficient to produce muscle contraction, but a *flow* of electricity most certainly sufficed. Could this be the force that drives life? If it was, then the body would have to make its own electricity. It could hardly rely on being plugged into an external source, in the manner of Frankenstein's monster.

Galvani set about searching for the "electric fluid" or "animal electricity" that living bodies would have to contain, and which his experiments had suggested might still be left as a residue in the dead animal. He had little success until his wife took a hand. The hand she took was one that she inadvertently put too close to the electrical generator while Galvani was performing one of his numerous experiments, drawing a spark that undoubtedly made her jump, but which also made the frog's legs jump, even though the current from the spark had passed through Galvani's wife and not through the frog. It seemed that it was the spark itself, and not the current, that caused the frog's muscles to contract. The most likely explanation, Galvani reasoned, was that the "electrical atmosphere" (we would now call it the electric field) from the spark was stimulating movement of "electric fluid" that was still contained in the frog's nerves.

There was still the disturbing possibility, however, that the "electrical atmosphere" from the spark was having a direct effect on the nerves and muscles of the frog. Galvani wrestled with the problem for six years, trying numerous experiments to try to distinguish between the two cases. One of his most spectacular experiments was to use lightning as the source of the spark. He took his prepared frog to the terrace of the Palazzo Zamboni in Bologna where he worked, placed it on a table with a wire running from the hook in the frog to a hook on the wall, and waited for a lightning flash. To his great delight, he observed that the legs kicked every time there was a distant lightning flash.

Figure 6.2. Frog in a lightning storm, plate 2 of Galvani's *Commentarius.*

The wire was acting as an aerial, with the electric field from the distant lightning flash inducing a small electric current, just as the field from a mobile telephone mast generates a small electric current in the aerial of a mobile telephone.

Galvani didn't know that, of course, but he did recognize that his prepared frog's legs were an extremely sensitive indicator of the presence of electricity in the atmosphere — "the most sensitive electrometer yet discovered," as he put it. But then came the fortuitous experiment that convinced him that animals really do make their own electricity. In his own words:

> In early September, at twilight, we placed . . . the frogs prepared in the usual manner horizontally over the [iron] railing. Their spinal cords were pierced by iron hooks, from which they were suspended. The hooks touched the iron bar [of the railing]. And, lo and behold, the frogs began to display spontaneous, irregular, and frequent movements.

It was the most exciting moment of Galvani's life. There was no lightning in the air, no sparks, and no external source of electricity. The only source of electricity to move the muscles must be within the nerves of the animal itself! In a flash of inspiration, Galvani saw the nerve as a biological Leyden jar, storing different charges on its internal and external surfaces, primed to make the muscle jump, in the same way an ordinary Leyden jar can make a line of people jump. Galvani's picture of the electrical state of the nerve was remarkably close to the modern one, although its mode of action is very different from what he envisioned, which was that an "electrical fluid" would flow through a nerve like the continuous flow of water through a pipe. That difference turned out to be crucial to the outcome of his forthcoming battle with Volta.

We now know that nerves generate and transmit electrical signals in a way that Galvani could not possibly have known about. The nerve cell (technically called a *neuron*) is a living cell with a long tubular extension (called an *axon*) down which the electrical signal passes. It is this axon that we commonly call a nerve. The axons in some animals (such as the squid) are so wide (around half a millimeter) that physiologists have been able to put electrical probes inside them and show that there is a voltage difference between the inside and the outside, just as Galvani had said.

When I visited the marine biology laboratories at Plymouth in England early in my career, I was able to watch some of the pioneering experiments with the squid axon firsthand. The axons were seven to ten centimeters in length, and I was surprised to find that their contents were not liquid, but gel-like. This did not impede the passage of an electrical signal, but it did make it difficult to insert the fragile glass-sheathed electrode into one end to make electrical contact with the inside. To get around the problem, the experimenters would roll a small squeegee along the axon, squeezing out a rod of jelly and making room for the electrode.

The electrode was used to measure changes in voltage after the axon had been stimulated at one end, producing an electrical pulse (called an *action potential*) that traveled to the other end at a speed of around 25 meters per second. The experiments that I was watching were designed to test the effect of drugs on the speed and frequency of the action potential. One of the results of such studies has been the discovery that alcohol affects our moods and mental abilities by making it harder for the nerve to produce an action potential. Despite their knowledge of this effect, the experimenters (and their guests) continued to enjoy a drop of Scotch whisky at the end of a long evening of experimentation. One nameless experimenter enjoyed his drop so much that, when he thought the time had come, he "accidentally" broke an electrode, finishing the experiment for the night.

Galvani described his results and conclusions in his 1791 book *De Viribus Electricitatis in Motu Musculari Commentarius* and sent a copy to Alessandro Volta. Like all Italian scientists, Galvani was in awe of Volta, a man who, by his own less-than-modest admission, had "a genius for electricity." At the age of thirty-nine he had already made many important contributions to the subject and had received prizes and awards for his work from many countries. Galvani was well aware that Volta's approval meant acceptance, and that his disapproval meant doom. What would Volta think of the proposal that there existed a totally new form of electricity?

Initially, the answer was "not much." Volta's first reaction to the notion of animal electricity was one of incredulity, but as he read and digested Galvani's book he shifted, in his own words "from incredulity to fanaticism." He said of Galvani's work that it was "proven," "certain," and that it "proves animal electricity with direct experiments and places it among the demonstrated truths." But when he tried to repeat Galvani's experiments for himself, he began to have doubts, especially when he found that he could elicit muscular contractions simply by touching the muscle and nerve with the tips of a pair of linked wires made from different materials. No external electrical generator was needed. To Volta, this meant only one thing — the wires themselves were producing the electricity, and the whole edifice of Galvani's argument fell to the ground.

Galvani fought back. After all, his original experiment had involved *similar* metals — an iron hook in contact with an iron railing. Volta argued that the iron of the railings and the iron in the hook must have had different compositions. Galvani responded with an experiment showing that contractions could be produced even when the conductors were

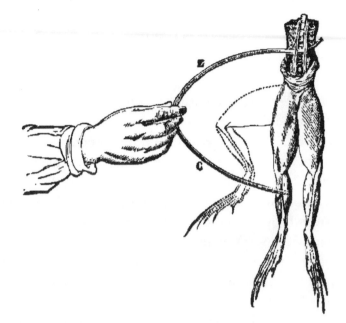

Figure 6.3. Frog's leg being stimulated by a pair of linked wires.
SOURCE: J. Malmivuo and R. Plonsey, *Bioelectromagnetism* (London: Oxford University Press, 1995). Internet version http://butler.cc.tut.fi/~malmivuo/bem/bembook/00/ti/htm.

nonmetallic. Volta replied by generalizing his theory to encompass all conductors, metallic or nonmetallic. And so it went on, with each side producing crucial experiments that seemed to support its own interpretation and to disprove the interpretation of the other camp. Soon the whole Italian scientific community was split:

> The storm aroused by the publication of the *commentaries* among physicists, physiologists and physicians [could] be compared only to the storm that at the same time arose on the political horizons of Europe. It [could] be said

wherever there were frogs, and wherever two dissimilar metals could be fastened together, people could convince themselves with their own eyes of the marvellous revival of severed limbs.

Volta was quite right. Dissimilar metals (or any conductors) in contact *do* produce their own electricity (this phenomenon, discovered by Volta, is known as the "contact potential"). But Galvani was also right. Animal nerves also produce their own electricity. Unfortunately, he didn't know how they do it, and this left a hole in his argument that Volta was quick to seize upon, and which led to the invention of the battery.

Volta's idea was simple. If he could replace the frog in Galvani's experiment by something inanimate, like a piece of wet cardboard, and still produce electricity by contacting it with a pair of dissimilar metals on either side, then that would mean that living (or recently dead) matter was not necessary for the production of the electricity. He described his experiment on March 20, 1800, in a letter to Sir Joseph Banks, secretary of the Royal Society in London. The metals (silver and zinc) and the cardboard were in the form of disks, about twenty-five millimeters in diameter. Volta multiplied their effect by stacking them as high as he could manage in a pile, in the order *silver, zinc, pasteboard, silver, zinc, pasteboard, silver, zinc, etc.* When he touched the two ends and received a shock, he knew that his pile was producing electricity.

The experiment had a huge effect on the Italian (and world) scientific community, which up until then had been divided between Galvani and Volta in its support. Its impact was such that Galvani's experiments and interpretations were utterly discounted from that time on. Volta used it to argue that "animal electricity" was a fiction, and that even its best-

known example, that of the *Torpedo* (or electric ray), func-
tioned in the same way as his battery. "Not even in this case,"
he said, "is it proper to speak of *animal electricity,* in the sense
of being produced or moved by a truly vital or organic action. . . . Rather, it is a simple physical, not physiological,
phenomenon — a direct effect of the Electromotive apparatus contained in the fish."

Volta argued that his experiments with different metals,
both on their own and in the form of his "pile," conclusively
showed that "herein lies the whole secret, the whole magic of
Galvanism. It is simply an artificial electricity, which acts under the impulse of different conductors."

Volta's discoveries of "artificial electricity" from the contact potential and the voltaic pile have found widespread application. Ironically, one of the less obvious applications is named
after Galvani. It is the process of *galvanizing,* where a sheet of
iron is covered with a protective layer of zinc. Many of the
houses in my Australian youth had roofs constructed of galvanized iron, and thanks to my father's early explanation of how
it worked I was the only one in my school class who was able
to answer the question when it came up — at least, I knew that
the zinc acted as a "sacrificial layer" that gradually corroded
away while the iron was left unharmed. Only later did I discover that it was the contact potential between the iron and the
zinc that was responsible for the gradual dissolution of the zinc.

The contact potential (mediated through a moist salty or
acidic layer between the metals) is also responsible for the
functioning of the battery, of which car batteries and flashlight batteries are obvious modern examples. A less obvious
example is the potato clock, where an electric clock is kept
going by means of a potato (no batteries required). All that
one does is to attach wires of different materials to the clock's

terminals and plunge them into a potato (or any other vegetable). The small voltage that is generated can keep an appropriately designed clock going indefinitely (or at least until the vegetable rots).

Contact potentials, though, were not always necessary to stimulate a nerve. Galvani's nephew Aldini had performed an experiment in which he had shown conclusively that dissimilar metals were *not* necessary for a dissected frog's leg muscles to be stimulated. All that was needed was a single piece of silver wire. When Aldini touched the nerve of a dissected frog with one end of the wire, and the muscle with the other end of the wire, the muscle contracted and the leg jumped. It was a result that flatly contradicted Volta's claims, but Volta was so convinced by the success of his own experiments that he resorted to the time-honored tactic of saying that there must have been "something wrong" with Aldini's experiment.

When I first read about Aldini's experiment, I too thought that there must have been something wrong. I couldn't see how a single piece of wire could generate an electrical signal, no matter how it was applied. Then I began to think more seriously about Volta's experiments with dissimilar metals, and soon realized that they shouldn't have worked either. The problem was that it requires around twenty-thousandths of a volt to initiate a nerve impulse, which doesn't sound like much, but contact potentials only produce around forty to sixty *millionths* of a volt. There didn't seem to be any way that a contact potential could initiate a nerve impulse in the manner described by Volta.

I called a few physiologist friends, expecting that they would be able to provide an answer off the cuff. None of them could (probably because I had failed to mention the presence of the iron hook!). When I mentioned the hook to Emeritus Professor Peter Barry at the University of New South Wales, he saw its significance immediately. He correctly pointed out the irrelevance of contact potentials, dissimilar metals, and the other factors that had plagued Volta and Galvani, and which were now plaguing me. All he was concerned with was that both Volta and Galvani had been poking huge pieces of metal (the point of a hook or the tip of a piece of wire) straight through the walls of delicate nerve cells, setting up a short circuit between the outside and the inside of the cell. As soon as the other end of the wire touched the tissue near a muscle, this completed a circuit that induced the nerve to fire.

I squirmed. Having seen the experiments on the squid axon at the Marine Biology Research Station in Plymouth, I *knew* what happened when the careless (or deliberate) movement of a tiny electrode disrupted the delicate membrane that surrounded the axon. The axon would invariably respond by firing electrical signals as rapidly as it could. I should have realized that exactly the same thing would have happened when the axon was impaled by a hook or a piece of wire.

Much of our knowledge of how the axon produces its electrical signals has come from studies on the squid axon. Some details of how an action potential is propagated along the axon are given in the notes to this chapter, but the initiating factor is a short circuit that is created not by poking a wire through the membrane or tearing a hole in it, but by a channel across the membrane that opens under an appropriate stimulus (such as a signal from an adjacent nerve or one of the body's many environmental sensors) to allow an inrush of positively charged sodium ions. Ironically, a good deal of information about the sodium channel has come from studies

on *Torpedo* and its electrically endowed counterparts, especially the electric eel (*Electrophorus electricus*) — the very group
of animals in which Volta used to argue that biological electricity did not exist. The electric organ of the eel contains a
very high concentration of sodium channels, which are mostly
responsible for its devastating ability to deliver a lethal charge.
Despite this drawback, the electric eel is still a favorite source
of sodium channels for laboratory studies.

The eel's life as a laboratory animal is quite a comfortable one, as I found when I visited the laboratory of a friend
in Denver, Colorado. The eels were kept in large tanks at a
comfortable temperature (for the eels), with prominent warn-
ing notices about the effects of carelessly putting one's hand
in the water. Laboratory assistants fed the eels at regular inter-
vals with tasty tidbits and would occasionally carelessly touch
the water in the tank. When they did, they received a belt
that was more than sufficient to remind them of the power of
natural electricity, and which no doubt brought a wry smile
to the face of the eel. If they had received a more prolonged
belt, as some people have when encountering the eels while
wading across a river, death could well have been the outcome.

The downside to the eel's blissful existence came when
some of the tail material was needed for an experiment. Anes-
thetic would be added to the water in the tank, and when the
eel had succumbed a small slice would be taken from the end
of its tail. The eels appeared to suffer no ill effects from this
procedure, but one could always tell which eels were new to
the laboratory and which had been there for a while, because
the longest-serving inhabitants were considerably shorter than
those that had newly arrived.

Electrically endowed fish such as the eel provided the first examples of electrotherapy. Scribonius Largus, court physician to Emperor Claudius, recommended the use of the black Atlantic torpedo ray, *Torpedo nobilan,* as a cure for gout:

> [A] live black torpedo should, when the pain begins, be placed under the feet. The patient must stand on a moist shore washed by the sea and should stay like this until his whole foot and leg up to the knee is numb. This takes away present pain and prevents pain from coming on.

A hundred years later the Roman physician Galen recommended the application of a live *Torpedo* to cure a headache. It seems to have worked, although I for one would not care to have *Torpedo*'s fifty or so volts sent between my ears. If the cures worked, they could possibly have done so through upsetting the nerve's ability to transmit a pain signal by means of a regular series of pulses. The more frequent the pulses (stimulated by a pain receptor), the more pain we experience, but the nerve cannot transmit more than a thousand or so pulses per second, and if it is forced to try to do more, the paradoxical result may be that the signals become scrambled or blocked and the experience of pain recedes. This principle is now being applied to the relief of arthritic pain by applying capsaicinoids (the active ingredients of raw chili powder) to the affected joint to "overirritate" the nerve. The efficacy of electrical stimulation, however, is much more open to question. The most widespread technique is TENS (transcutaneous electrical nerve stimulation), which is based on the principle of using electrical stimulation to control the "gating currents" that open or close channels in selected nerves. TENS is used by nearly half a million people annually in Canadian state hospitals alone, and is widely available in

Canada and the U.K. for the treatment of labor pain. Despite its widespread usage, the results of clinical trials to date have shown that "there is no evidence that TENS provides effective pain relief" for either chronic pain or labor pain. The technique still has many advocates in the medical profession, however, and it may be that some modified form will be discovered that does have a real effect.

Electrotherapy for the restoration of movement to paralyzed muscles has, as mentioned earlier, an even less impressive history. There have been some well-attested exceptions, however. One of the most striking was the case of a Swiss locksmith named Nogues, who in 1733 suffered a blow to the head that rendered him almost completely paralyzed on the right side. Fourteen years later, he came under the care of Jean Jallabert, Professor of Experimental Philosophy at Geneva. Jallabert had been performing experiments with the newly invented Leyden jar and decided (with the assistance of Daniel Guiot, Geneva's leading surgeon) to try its effect on Nogues's arm. He warmed the arm and administered shocks to the now withered limb for an hour and a half each day. At the end of a month, Guiot could lift a full glass of water, and by the end of three months, with the aid of a course of exercises, he had completely recovered use of the arm.

This long-term, careful approach to the electrical stimulation of muscles had its successes, but such electric treatments largely remained the province of quacks and only began to achieve respectability after the 1871 publication in America of *Medical and Surgical Electricity* by George Beard and Alphonse D. Rockwell. Even then, it was another six years before the New York Medical Society allowed the authors to discuss the subject in its meetings. It is now an accepted part of the therapist's armory. It is also, unfortunately, the territory of much modern quackery. There is one area, however, where quacks have not dared to trespass, and where real scientific medicine has made enormous strides. That is the area of the heart.

If the heart stops, death follows, unless the heart can be restarted pretty rapidly. The first person to show that it was possible to do the job electrically was (you guessed it) Aldini, who publicly suffocated a dog to the point of cardiac stand-still, and then resuscitated it with thoracic shocks from a voltaic pile.

Some fifteen years later, the American doctor Richard Reece published a family medical guide that included a won-derful description of "The Animation Chair of Doctor De Sanctis," in which a patient could be resuscitated with equip-ment that included a bellows with laryngeal tube to inflate the lungs, a heated globe to create inhalant vapors, and a voltaic pile with a silver tube (from one electrode) that was passed into the gullet, while a wire from the other electrode was "successively made to touch different parts of the external surface of the body, particularly about the regions of the heart, the diaphragm, and the stomach during the inflation of the lungs." "This is little different," says the expert Ellen Kuh-feld, curator at the Bakken Library and Museum of the His-tory of Electricity and Medicine, "from defibrillation and external pacing as they are practiced in modern medicine."

"External pacing" means applying intermittent electri-cal impulses to the heart muscles at just the right frequency to keep those muscles beating regularly. It was invented by the American surgeon Albert Hyman in 1932, and it had a suc-cess rate of about 30 percent when used during surgery. Even this success rate was regarded by some as an affront to the Almighty, and "Hyman was beset with abusive correspondence, and even lawsuits, from irascible people who regarded his resuscitation endeavours as sacrilegious tampering with Divine Providence."

The main danger to patients, though, was not divine providence but the power companies. When surgeon C. Walton Lillehei at the University of Minnesota began in the 1950s to insert stainless steel wires into the heart before closing the chest after a heart operation, he used the power mains as a source of electricity. A young patient died after a power failure, initiating a quest to invent a permanently inserted, battery-driven pacemaker. The success of the quest is well-known. Pacemakers were developed that could be left in the body and gradually became smaller and more reliable, thanks partly to the invention of miniaturized batteries and other components, and also to our increased understanding of just how biological electricity works. The wheel has thus come full circle, and Volta's invention, intended to disprove the existence of biological electricity, is now being used to keep patients alive by simulating, and sometimes stimulating, the very source of electricity whose existence it was intended to deny. We may not be able to create life, in the manner attempted by Dr. Frankenstein, but as a result of Galvani's insights and the inventions of Volta, we now have a much better chance of keeping it going.

7

What Is Life?

This book began with the story of Duncan MacDougall, the American doctor who controversially believed that he had weighed the human soul. It ends with the story of another controversy about the nature of life — the argument between the "mechanists" and the "vitalists" about whether life consists only in the functioning of organic beings according to physical laws, or whether living beings are additionally imbued with some sort of "vital force."

The controversy goes back to Aristotle, but it received renewed impetus with the result of an experiment to manufacture an animal clone (a hundred years before Dolly!). The embryologist Wilhelm Roux, representative of the new breed of mechanists, had concluded from his experiments on the development of frog embryos that the processes of life could be accounted for in purely physical terms. The respected zoologist Hans Driesch, however, concluded from an experiment in which he artificially produced a clone of genetically identical sea urchin embryos that living organisms must contain a "vital force" peculiar to life itself. He developed his ideas into a theory known as vitalism, which split the biological community, and which took biological science back nearly a century to the days when the prevalent view was that nearly all life phenomena were guided by an all-pervading vital force.

Vitalism was a powerful movement indeed. Even as late

as 1930 the eminent historian Emmanuel Radl was able to write that "Driesch marks the end of Darwinism." He didn't, of course, but the hidden hand of Driesch can still be felt today in arguments over cloning, in vitro fertilization, abortion, and man's nature and place in the universe. A reexamination of Driesch's evidence showed that it actually provided strong support for the mechanists' views. This chapter describes how the interpretation of Driesch's evidence was transformed and shows that his real legacy to science was to force a painful reexamination of what science can and cannot tell us about the nature of life and about how such complex arrangements of cells as ourselves can develop from the division of a single fertilized egg.

"Where did I come from?" is a question that most of us have asked as children, and which led to further questions about the nature of our existence as we grew older. When I first asked it of my mother, she turned bright pink and said that she would buy me a book to read. If the young Aristotle had asked the same question of his mother nearly 2,500 years ago, she would not have known the answer and would not have had any books to guide her precocious young son. It was Aristotle who wrote the first book on the subject. In it, he proposed the idea that the woman supplied the matter for the body of the developing embryo (from her "monthly courses") and that the role of the male was to provide an animating principle, or soul, to activate the material substance and to provide it with life.

This flattering picture (from the male point of view) held sway for nearly two thousand years. Even when individual eggs and spermatozoa were first observed through the microscope in the sixteenth century, the roles that were ascribed to them followed those ascribed by Aristotle. The big question was not whether Aristotle was right but just *how* the soul provided form. One popular guess was *preformation* — the idea that the sperm contained within it all future generations in miniature, enclosed within each other in the manner of a set of Russian dolls. The dolls were called homunculi, and proponents of the view that they were contained within the sperm were buoyed up when, shortly after the invention of the microscope, the Dutch microscopist Anton van Leeuwenhoek reported seeing homunculi in the heads of living spermatozoa.

I have to say that I sympathize with Leeuwenhoek, especially since the time when I was examining the head of a spermatozoon through a powerful microscope and found the face of British prime minister Tony Blair smiling back at me, formed by the pattern of wrinkles and bumps on the cell surface. It is all too easy to see things that aren't there when looking at tiny objects under the microscope, and Leeuwenhoek's microscope was a primitive one, with low resolving power. Nevertheless, his observations were taken by the "spermists" as triumphant confirmation that they were right and that the "ovists," who believed that homunculi existed but that they were to be found in the egg, were wrong.

The battle between the ovists and the spermists is recounted by Clara Pinto-Correia in her book *The Ovary of Eve*, with the title making the point that homunculi, if they existed, must have originated with Adam and Eve. Aristotle would probably have disagreed with both sides. He was in favor of *epigenesis* — the new formation of something organized out of that which is devoid of order.

By the middle of the nineteenth century, it was becoming

obvious to embryologists that Aristotle had been closer to the mark than those who believed in preformation. The development of multicelled animals from a single fertilized egg seemed to follow a regular pattern, no matter whether the egg was that of a bird, a frog, a sea urchin, or even a human. The egg cell would divide into two, then four, then eight, and would eventually become a hollow ball of cells (technically, a *blastula*). Then the ball would start to fold in on itself (forming a *gastrula*), with the infolding eventually becoming the interior of the gut. The several layers of cells thus formed would proceed to *differentiate* to perform different functions, with the outer layer forming skin, hair, etc., the middle layer forming muscle and nerves, and the inner layer forming the digestive tract.

The details of this true miracle of life were teased out in the latter half of the nineteenth century, at a time when scientists were becoming convinced that many, if not all, of the mysteries of life could resolve themselves into matters of physics and chemistry. They had some reason for their belief. It had already been shown that urea, an "organic" substance thought to be produced only in the urine of living animals, could also be manufactured from materials that were in no sense living. It had also been found that the fundamental physical principle of the conservation of energy, which says that a certain amount of work produces an exactly equivalent amount of heat, could be applied to living beings just as it could be applied to machines. There seemed to be no reason

Figure 7.1. The miracle of gastrulation.
A: Starfish embryo containing twelve cells, with one in process of division. *B:* Starfish blastula (hollow ball of cells). *C:* Starfish blastula in process of infolding to produce gastrula. PHOTO: Barbara Payne, Cuyahoga Community College

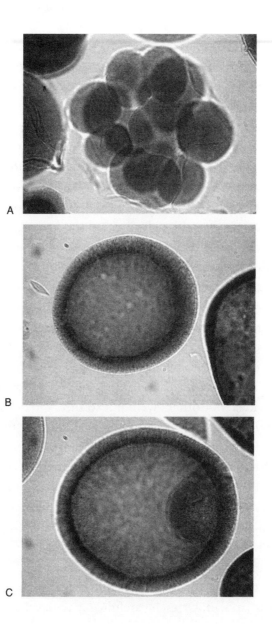

why life should not be based on some very complicated form of machine.

One who believed this philosophy was Wilhelm Roux, an embryologist who worked in Breslau (now Wroclaw), situated on the river Oder in modern-day Poland. Roux firmly believed that the development of a living organism from a single fertilized egg was controlled by an internal mechanism. Ruthlessly following his own logic, he saw that this would only be possible if the machine divided each time the cell divided, so that the first generation of daughter cells would contain half of the machine each, while the next generation would contain a quarter each, and so on. The eggs that he chose to study were those of the frogs that were in plentiful supply in the freshwater streams around Breslau. He conceived the brilliant experiment of taking a fertilized egg after it had divided to produce two daughter cells and killing one of the cells by piercing it with a hot needle. To his immense gratification, the other cell continued to divide but to produce only half of the components necessary to make a complete frog.

It seemed that Roux had dramatically confirmed the "machine" idea, but others were not so sure. One of the unbelievers was Hans Driesch, a zoologist working at the International Zoological Laboratory on the shores of the Bay of Naples. Driesch was already famous for an experiment in which he cut planarian worms in half and then rejoined the halves so as to create a living, moving planarian with two heads or two tails. The experiment required delicate manipulative skills, which he put to good use when he began to study the development of sea urchins, which were as plentiful on the floor of the Bay of Naples as frogs were in the streams near Breslau. Driesch carefully separated the daughter cells of a sea urchin embryo that had developed to the "four-cell" stage, and gave each cell a chance to develop independently. He found that the daughter cells developed, not into the one-quarter of a sea urchin that Roux would have predicted, but into a complete (albeit small) adult. This experiment seemed to prove that the sort of machinery proposed by Roux could not exist since, in order to develop into a whole animal, each daughter cell would have had to contain *all* of the machinery contained in the original fertilized cell. "*How could a machine be divided . . . ,*" argued Driesch, "*and yet remain what it was?*"

Driesch also had logic on his side, since a moment's reflection shows that any machine that is repeatedly split into pieces must eventually end up in pieces so small as to be useless. His next logical step, however, split the scientific community. He concluded that, since he could not visualize how a machine could drive a living cell, then cells must be driven by something else, outside the normal laws of physics. He called it vitalism, and with the invention of this name in 1892 the battle lines between the mechanists and the vitalists were drawn.

Driesch chose a fortunate time to instigate his battle. The success of the mechanists in explaining some aspects of biological function was seriously upsetting some biologists, who argued (as some biologists still do today) that living processes were too complicated to analyze in physical terms. The mechanists were also not making things any easier for themselves by adopting an extraordinary hubris. Some were so convinced that all life processes would eventually be explained in purely physical terms that they were ready to advance the most outrageous theories along these lines, including the notion that ideas must be made of a physical material that is secreted by the nerve cells in the same way that urine is secreted by kidney cells.

The stage was thus set for a battle between the mechanists and those biologists who wanted to take the mechanists down a peg or two. Vitalism provided a focus for the battle, which was initially concentrated on whether the results of Roux or Driesch were the more believable. The resolution came in an unexpected way when a mechanist produced the clinching evidence — but in Driesch's favor. The mechanist was the great Hans Spemann, founding father of modern developmental biology, who discovered that the development of the living daughter cell in Roux's experiment had been restricted by the presence of its dead sister. When the experiment was repeated, but with the two living cells being completely separated initially instead of one being killed off, both cells developed into complete organisms.

The resolution of the disagreement provided a new stimulus for vitalism. Driesch's experimental result was vindicated, and none of the mechanists had an answer as to how a machine could keep dividing (not just in the development of an individual, but also from generation to generation) and yet remain as it was. It was to be fifty years before the physicist Erwin Schrödinger showed that there was another possibility in his remarkable little book *What Is Life?*

Schrödinger is famous as the originator of wave mechanics, in which a single equation is used to describe the whole of the physical world in quantum mechanical terms. In *What Is Life?* he applied his physical insight to the biological world and showed that dividing cells in a developing organism must pass on to their daughters the *instructions* for building an internal cellular machine, rather than parts of the machine itself.

Schrödinger's book stimulated Francis Crick to move from physics to biology, where he famously participated in working out the structure of DNA and unraveling the genetic code. I read Schrödinger's book only after I had made the move to biology from the physical sciences, but I can still remember the frisson it gave me, as it does today when I reread it. Schrödinger proves, with a clarity and sharpness of argument that I can only envy, that the simple principles of physics inevitably lead to the conclusion that the self-replication and functioning of individual cells *must* be guided by a single molecule (or pair of molecules) in each cell, and that this molecule can occupy a space of no more than 30 nanometers cubed, or about a millionth of the total volume of the cell, which contains a total of some *billion billion* molecules, all of whose fates are ultimately guided by just the one molecule!

Schrödinger himself was stunned by his conclusion and did not "expect that any detailed information on this question" (about the nature of the guiding molecule and its mode of action) was "likely to come from physics in the near future." He also believed that "living matter, while not eluding the 'laws of physics' as established up to date, [was] likely to involve other 'laws of physics' hitherto unknown" — a conclusion that Driesch had also drawn, albeit from very different premises. In the event, both Schrödinger and Driesch were proved wrong. It was just eight years after the publication of Schrödinger's book that Jim Watson (stimulated by hearing a talk at the Naples Zoology Station where Driesch had worked fifty years earlier) with Francis Crick worked out the basic structure of DNA, the molecule whose existence Schrödinger had presciently forecast, and showed that its function required no new laws of physics or chemistry.

DNA fulfilled all of Schrödinger's criteria and in some

ways exceeded them. It seems impossible that a molecule wrapped up in such a tiny space could do the job, but the space is an illusion. If all of the DNA in your body were laid end to end, it would stretch from Earth to the Sun and back eight and a half times, and it would take light an hour to get from one end to the other. Plenty of room there for an instructional code, however complex!

Schrödinger does not quote Driesch's experiment in his book (in fact, he does not give any biological references at all), but Driesch's experiments are often quoted in modern textbooks as the first clear evidence that living cells must pass an instructional code unchanged to their daughter cells as they divide and begin to differentiate. His experiments are also cited as the first example of the deliberate creation of a clone of genetically identical animals. When it comes to plants, man has artificially produced clones since at least Roman times. When the Romans invaded Britain, they brought with them the branches and twigs of fruiting plants, which they grafted onto the root stocks of indigenous plants. The root stocks provided nutrients for the grafted branches, which grew to produce the desired fruit. In due course pieces of the fresh living branch would be cut off and grafted on to a new root stock. All of the new branches, and the new fruit, were genetically identical to those of the original plant (i.e., cloned), since they had developed from mature tissue cells rather than from a fertilized ovule (the plant equivalent of an egg).

Cloning is now a common practice among horticulturists. Flower growers use it so that their flowers can be guaranteed to look just like those produced by the parent plant. Even the fruit that you buy in the supermarket is likely to have been cloned — strawberries that have been grown from runners, for example, or Granny Smith apples, every single one of which was cloned from the original mutant, discovered by the Australian grandmother Marie Ana Smith in her back garden one hundred and forty years ago.

Around half of the plants in the world are natural clones, since most plants follow an "alternation of generations." One generation develops from the fertilized ovule, which grows into a plant that produces spores genetically identical to the parent, and which in turn grow into a plant that produces the ovules and/or pollen necessary for sexual reproduction. Examples of natural clones are rarer in the animal kingdom, but they are still there. Identical twins are clones, as are some drones in a hive of bees or the female workers in a colony of ants. Some female beetles, such as Fuller's rose weevil, also produce clones of offspring without the aid of a male, and there is even a lizard (the aptly named virgin whiptail lizard) that produces clones.

In general, however, animals can only produce clones of offspring with the deliberate intervention of people. In this respect, Driesch's experiment seems to have been the first example of the deliberate production of an *animal* clone, and he was certainly correct in claiming that it disproved Roux's idea of a cellular machine that divides each time the cell divides. His conclusion that a special life force must therefore exist was another matter altogether, and his arguments led to a dramatic reevaluation of what science can and cannot tell us about the processes of life. The reevaluation began at a time (around 1911) when the claims of the vitalists were becoming every bit as extreme as those of their mechanist rivals. The ardent vitalist R. Neumeister even argued, "The [gut] epithelium possesses as much sensation, as much judicial power to know what is good for the body, as the nerve cells of the cortex."

It would certainly make life a lot easier, and also destroy

the fad diet industry, if we had thinking guts that would re-ject any unsuitable food that we ate. It is true that our guts do reject some foods by mechanical or chemical processes, but Neumeister's argument for the existence of a thinking gut had the same flaw as Driesch's argument for a "vital force" — the assumption that, if we can't understand how something works, then its operation must involve forces beyond our un-derstanding. It is an argument that has led people in different parts of the world to join cargo cults and to believe in the spoon-bending powers of Uri Geller, the spontaneous gener-ation of living maggots from rotting meat — and the exis-tence of a special life force in living cells.

A major shot in the long-running battle against Driesch's logic was fired in a lecture by S. J. Melzer at the University of Buffalo in December 1904. Having been in Buffalo at that time of year, I can assert with confidence that the lecture hall would have been surrounded by several feet of snow, that the temperature would have been well below freezing, and that the lecture audience must have been hardy souls indeed. They were rewarded, however, with an important critical analysis of Driesch's logic (in a lecture whose topic was supposed to be "Oedema"!). Melzer argued that Driesch's vitalism was no more than a storage place for biological facts that were not yet otherwise explained. He called it "transcendental" vitalism, since it could never be proved or disproved. Then he produced a real shock by saying that there was *still* a role for vitalism, since it might happen that biologists would come across new phenomena that could *never* be explained by physical theories, and which could *only* be accounted for if cells contained a "vital force." He called it "natural" vitalism, since its truth would be subject to scientific test.

Melzer's analysis seems to me to have been devastatingly sensible. Perhaps that was why it was ignored at the time, and why proponents from the opposing sides continued to shoot from the hip in seminars and at zoological conferences, where vitalism was often reported to have been the main topic of conversation. The battle culminated in a series of very public and increasingly caustic exchanges between Arthur Lovejoy and H. S. Jennings, both from Johns Hopkins University, in which Lovejoy saw some forms of vitalism as fitting within the framework of normal science, while Jennings would have none of it. Their arguments, which were published in the journal *Science* between 1911 and 1914, led nowhere, because the combatants were fighting on two different fronts, but they did bring the problem to a focus, and confirmed that whether vitalism was scientific or not depended on what you meant by vitalism — just as Melzer had said.

Melzer's natural vitalism has a historical precedent within the hallowed halls of physics. Until the 1850s, it seemed inconceivable that light, heat, and motion could just be different aspects of the same thing — a thing we now call energy. Nobody has ever seen, or measured, or felt energy itself, but all scientists now believe in its existence because, by using the concept, they can make greater sense of the things that they *can* see, feel, and measure.

Scientists were forced to accept the existence of energy when all other explanations failed. Vitalism (of the natural kind) may yet prove to fall in the same category, and it could still be needed to explain the apparently purposive behavior of living cells and the development of complicated multicellular organisms such as ourselves from a single egg cell. At the moment, however, most of the evidence seems to be pointing toward more mechanical explanations. The true scientific legacy of Roux and Driesch is that both of them unwittingly made major contributions to our understanding of how such mechanisms might work.

The Legacy of Roux

Wilhelm Roux's mistake may have been one of the most important in the history of biology. By leaving a dead cell attached to its still-living relative, he accidentally showed that the dead cell could profoundly influence the development of its living neighbor. It was the first indication of the importance of "nearest-neighbor" interactions in biology and particularly in the development of the embryo, where many biologists now believe they are paramount.

I became excited about nearest-neighbor interactions when I came across a paper in which the Princeton biologist Malcolm Steinberg suggested that different types of cell might have different "adhesivities," with those of similar adhesivity preferentially sticking together. To prove his point, he separated the cells from a chick embryo, shook them up in saline solution, and found that they stuck together in a ball, but with the muscle cells at the center, enveloped successively by layers of skin, heart, liver, and retinal cells. Steinberg's experiment showed that nearest-neighbor interactions between like cells can produce ordered patterns when the cells aggregate, and the underlying suggestion was that these were at the forefront when it came to the way that living cells organize themselves into the patterns that become nerve, muscle, skin, etc., as an embryo develops into a mature organism.

I was studying the adhesiveness of small mineral particles at the time, and I became hugely excited by the idea that I might be able to adapt the techniques I had developed to what sounded like an important biological question. The only problem was that I didn't know any biology. The subject had not been taught in Australian boys' schools in my day. It was regarded as a girls' subject, with the additional danger that boys might learn too much about sex if they were exposed to it. When I entered Sydney University it was to study chemistry, physics, and mathematics, and I learned too late of the fascinating opportunities that biology offered. When I came across Steinberg's "differential adhesion hypothesis," however, I knew that I had to go back to college and complete my education by learning biology.

It was the fulfillment of a childhood dream, originally stimulated by one of my favorite childhood books, *The Water Babies* by Charles Kingsley. I especially liked the picture of two scientists examining a bottled water baby with a magnifying glass, but when I used my father's glasses to try to see water babies for myself in jars of muddy water taken from nearby streams and puddles, I couldn't find any. The two scientists caricatured in the picture were real enough, though, and their different approaches to biology corresponded closely with those of the opposing factions in the vitalist/mechanist debate that was to come. One was the Victorian biologist T. H. Huxley, Charles Darwin's most famous supporter. The other was Richard Owen, a virulent opponent of Darwin and of the whole notion that life could be understood as any sort of mechanical process. Kingsley was a clergyman who sided with Owen, and the picture that entertained and stimulated me as a child was really a sideswipe at the whole idea of trying to understand life in scientific terms.

I was not the only budding scientist to have been stimulated by Kingsley's picture. Julian Huxley, T. H. Huxley's grandson, wrote to his grandfather at the age of five asking whether it was true that he had seen a baby in a bottle. His grandfather wrote back gravely:

My dear Julian,

I never could make sure about that water baby. I have seen Babies in water and Babies in bottles; but the Baby in the water was not in the bottle and the Baby in the bottle was not in the water. My friend who wrote the story of the Water Baby was a very kind man and very clever. Perhaps he thought I could see as much in the water as he did. There are some people who see a great deal and some who see very little in the same things.

When you grow up I dare say you will be one of the great-deal seers and see things more wonderful than Water Babies where other folks can see nothing.

Julian Huxley did grow up to become a famous zoologist, and the knowledge that they were teaching the precocious grandson of the great T. H. Huxley must have caused a certain amount of consternation among his lecturers at Cambridge. My own presence as a mature student in lecture classes at Macquarie University in Sydney also worried some lecturers, who were concerned that their sketchy knowledge of the physical sciences might be exposed. They needn't have worried. I was there to learn, not to score points, and in any case they soon discovered that they were teaching an ignoramus when it came to biological subjects. I can still remember the look of incredulity on the face of one lecturer when he found I had no idea that plant cells were different from animal cells and I had never seen either of them under the microscope. His incredulity was matched by my excitement at discovering a whole new world of knowledge. I especially enjoyed touring around the research groups in the school, finding out what each was doing and trying to understand what the important biological questions were. I soon became involved with one of the groups, led by Professor Keith Williams, and became fascinated by the behavior of the little single-celled amoeba *Dictyostelium discoideum* that he was studying.

Dicty (to give it the shortened name that everyone used in the lab) is a tiny amoeba that inhabits garden soil. Every shovelful of soil that you dig up from your garden probably contains hundreds of thousands of *Dicty* and its relatives. They are among the most valuable inhabitants of any garden, feeding on soil bacteria, controlling the bacterial population, and helping to replenish soil nutrients. As the amoebae feed, they grow, eventually reaching a size where they divide to produce two new amoebae, each as hungry as the original.

The problem for *Dicty* comes when the growing population begins to run out of bacteria to feed on. The individual amoebae, so small as to be invisible to the human eye, are quite incapable of traveling the large distances needed to get to a fresh source of food on their own. What they do to overcome the problem requires a degree of cooperation that beggars belief. When food supplies dwindle, the amoebae stream to a common meeting point and assemble themselves into a miniature slug, about a millimeter long and containing around 100,000 individual cells. This slug (technically known as a *grex*) uses the combined power of its individual members to pull itself up to the soil surface, where it erects a stalk that has a "head" loaded with spores, each one containing a living amoeba. When the head bursts, the spores are scattered by the breeze, and the amoebae are released to get on with the job of consuming more bacteria until starvation sets in once again.

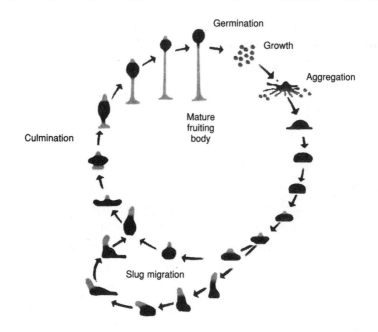

Figure 7.2: *Dicty* **life cycle.**
SOURCE: Professor Steven Alexander, University of Missouri, redrawn from *Trends in Cell Biology* 10 (2000), 215–19.

The walls of Keith's laboratory were covered with cartoons of him (with his very cartoonable wiry red beard) grappling with the mysteries of *grex* movement. The originator of the cartoons was Phil Vardy, Keith's anarchic Ph.D. student. I joined in the hunt with enthusiasm, throwing in ideas from a physical scientist's perspective and reveling in the knowledge that Keith's interest in *grex* movement had no relevance whatever to my official work as a food scientist in the government laboratory two kilometers down the road.

Phil's laboratory bench was covered with petri dishes containing *Dicty* at various stages in the life cycle that culminated in the *grex*. He showed me that the apparent deliberateness with which the amoebae moved to help their starving companions was the result of a very simple mechanism, whereby the starving amoebae released a chemical that attracted their nearest neighbors. The neighbors would respond by deforming so as to stretch out in the direction from which the chemical signal had come and pull the amoeba in that direction. Amoebae that were so stimulated would also release their own chemical signal, enticing other amoebae from progressively greater distances, until eventually streams of amoebae would be moving toward the original source of the signal. This behavior could be induced simply by placing a small amount of the chemical signal itself in the middle of the dish. There was no need for a teleological explanation at all.

The fascinating thing about the movement of the amoebae was the pattern they formed as they moved, which consisted of a set of well-defined rings that moved like battalions of soldiers attacking a citadel. It was my first introduction to the idea that purely local interactions can create complex patterns. One of Keith's students was even able to replicate the patterns by computer simulation. Each amoeba was represented by a dot on the screen, and each dot was programmed to move toward any neighbor that was "emitting" a signal, and to emit its own signal as it went. As soon as one dot in the middle was instructed to emit its signal, concentric rings of dots converging toward a central point quickly formed on the screen.

Simple patterns such as those formed by *Dicty* are one thing; the patterns formed by the cells in the human body are quite another. Many early biologists (and some present-day ones) claimed that bodily organs like eyes, kidneys, and livers need such complex arrangements of cells that simple physical and chemical interactions alone could not possibly account for their formation. When I took time off from my biology degree to become a visiting scientist at the Middlesex Hospital Medical School in London, I discovered that quite the opposite was true, and that even the formation of legs or wings from the cells of a vertebrate embryo may be the result of simple physical or chemical cues.

The dank, smelly corridors and indescribably grubby tea room in the Anatomy Department gave no hint of the wonderful work that was going on behind its closed doors. One of those doors belonged to Lewis Wolpert, who was testing an idea first proposed by Alan Turing, the father of modern computing and linchpin in the breaking of the Enigma code during the Second World War. After the war Turing developed an interest in the more complicated code by which living cells choose how and where to specialize during development and suggested that cells might use information about their position within an organism to decide what specialism to adopt. It seems impossible that a cell could "know" where it is, but Turing suggested that cells may be able to detect and respond to the concentration of a chemical released by a distant neighbor, since the concentration of the chemical would diminish with the distance from its origin. If several sources of different chemicals were available, cells might respond to the ratio of the concentrations to get accurate information about their positions.

It was a clever idea, which Turing reinforced by computer models showing how the stripes on a zebra or the spots on a leopard might develop in this way. Lewis decided to test Turing's hypothesis in real animals and showed that the developing cells in embryonic chick limbs use just the sort of "positional coding" that Turing had proposed. I was not involved in this work but was often confronted with the unnerving situation of having Lewis stride into my office, demand the answer to a mathematical question, and stride out again without breaking pace if the answer was not immediately forthcoming.

Lewis showed that the chemicals released by distant neighbors could affect the developmental fate of a cell, but the overwhelming weight of evidence now shows that it is the chemicals released by a cell's closest neighbors, or even the mere physical presence of those neighbors, that determines the developmental fate of many cells. The first person to recognize this fact was Hans Spemann, the man who showed where Roux had gone wrong, and who revealed the precious clue hidden in Roux's work.

Spemann's great passion was Nature, and he loved to walk in the countryside around Dahlem on the outskirts of Berlin where he worked. The best months for walking were in the summer, but these were denied to him, because summer was the breeding season for newts, and Spemann spent every summer for forty years in his laboratory studying the development of newt embryos. He had discovered where Roux had gone wrong by repeating Roux's experiment with these embryos, and in doing so had gained the first clue about the importance of nearest-neighbor interactions in development. Inspired by his success, he decided to test the effect of separating nearest neighbors in the cleavage stages of newt embryos by using the fine blond hairs of his baby daughter to constrict the embryo so that a bottleneck remained between the two halves. Often, the newt would become two newts, joined at the trunk in the manner of conjoined twins:

Spemann's experiments caused huge public interest and

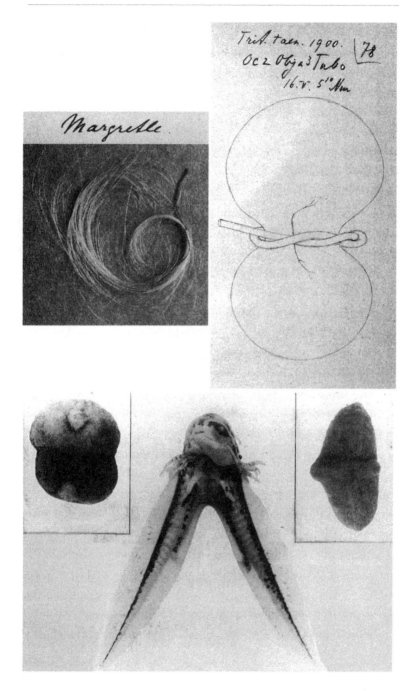

Figure 7.3. *Top left,* **blond hairs from Spemann's daughter Margrette, found in Spemann's laboratory files after his death.** *Top right,* **sketch by Spemann of "bottleneck" embryo.** *Bottom,* **"conjoined twin" newts.**
SOURCE: K. Sander and P. Faessler, *International Journal of Development Biology* 45 (2001): 1–11.

led to speculation about whether two souls could be created out of one if the same procedure were to be applied to a human embryo. Spemann, meanwhile, had noticed that two separate embryos developed only if the original embryo was divided in such a way that a particular group of cells was split between the two halves. His student Hilde Proescholdt followed up this clue in 1921 by transplanting a few cells from the area of an embryo where the head and spinal cord were beginning to develop to the opposite side of a different embryo. To her amazement, she ended up with an embryo that had two heads and two spinal cords. She had discovered the "organizer region" — a small group of cells whose intimate presence totally controls the course of development of those around them. Tragically, Hilde did not live to reap the rewards of her remarkable experiment, which led to the award of the Nobel Prize to Spemann in 1936. Shortly after the work was finished and published in 1924, she died in a kitchen fire caused by an exploding stove.

The evidence that the intimate presence of one cell or a small group of cells can affect the developmental fate of their neighbors is now overwhelming. Evidence is also accumulating that nearest-neighbor interactions control how cells arrange themselves into patterns. The process is called *self-organization,* and some of the ways in which it happens are among the most remarkable in nature. Almost all of them rely on very simple rules, which can produce astonishingly complex patterns.

One of the simplest examples is the cooperative movement of a shoal of fish, where the rule is "do what your neighbor does." When a migrating shoal of sardines is attacked by a predator such as a dolphin, for example, each sardine follows the movement of its nearest neighbor, but at such a speed that the whole shoal seems to be ducking, dodging, and flashing its multiple silver bellies as if a "supermind" were guiding its actions. High-speed filming, however, supported by computer modeling, reveals that each sardine is simply mimicking the movements of its nearest neighbor. Another example is in the behavior of near-blind army ants, which may not form the two-mile-wide, twenty-mile-long columns found in Hollywood films such as *The Naked Jungle,* but which certainly form into columnar raiding parties containing some 200,000 members, whose efficient organization requires only two simple rules: follow the odor trail left by the ones in front of you, and turn to the side if you meet an ant coming the other way. As a result of these rules, ants returning to the nest with prey find themselves moving down the center lane of a three-lane highway, with those as yet unburdened (and hence able to turn more quickly) moving to the outside lanes.

When individual living cells follow simple rules in response to their nearest neighbors, they too can form complex patterns. Some of those patterns arise when cells migrate by following trails laid down by others, just as army ants do. Even a difference in stickiness can be enough to cause cells to group together in an organized way, just as Steinberg discovered. My own work on the stickiness of living cells took a different direction, in which I followed up the observation that cancerous cells are less sticky than normal cells, and hence more liable to come loose from tumors and travel to other parts of the body where they lodge and form secondary cancers. Scientists who were working on normal development found that many cells are coated with selective "adhesion molecules," and that they also respond to a wide range of chemicals that may be emitted by neighboring cells. The local messages are thus there (perhaps disrupted in the case of cancer), and a hugely surprising, counterintuitive conclusion has emerged from the mathematical analysis of computer simulations about how cells might use those messages. It is that "purely local interactions can generate arbitrarily complex spatio-temporal patterns." This phrase, stripped of its jargon, means that *it is possible that all of the patterns and movements in living organisms may come down to how individual cells respond to their nearest neighbors.* This is not to say that nearest-neighbor interactions *do* determine how the pattern of cells in our bodies arises, but it does say that there is no reason why they couldn't.

Driesch's Legacy

Once the pattern is formed, the question remains as to how the cells in different positions differentiate to perform different functions. Driesch's experiment with sea urchin embryos provided the first vital clue as to how this might happen. It proved that if cells do contain internal machinery that directs their development, then that same machinery must be passed on to the daughter cells *in toto* when the parent cell divides. Driesch couldn't see how they did it and was forced to postulate that the machine did not exist, and that cells were controlled by a "vital force." We now know that the machine *does* exist, and that it perpetuates itself by passing the *information* to the daughter cells on how to make fresh copies of itself. The machine is called DNA. Most people will have seen images of the two strands of DNA in a parent cell unwrapping, with one going to each of the daughter cells and there acting as a template for the formation of a new second strand from chemicals existing within the cell.

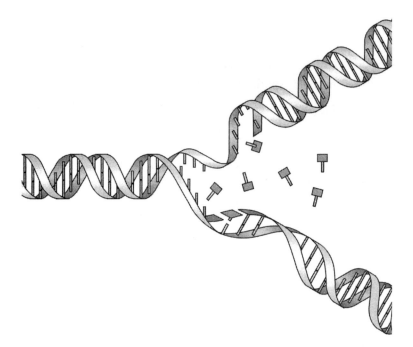

Figure 7.4. Double-stranded DNA unraveling.

The DNA forms the *genome* — a series of coded instructions that tell the cell what proteins to manufacture to do its job. All cells contain the whole genome. The difference between cells that do different jobs is that different parts of the genome are activated, so that the cells are instructed to manufacture different proteins. The totality of proteins that a particular cell manufactures is called its *proteome* — a term coined by Mark Ilkins of Keith Williams's laboratory as a result of a competition to name this new field and now used around the world. The proteome for some cells on the surface of the body consists almost entirely of keratin, which forms our nails and hair. The proteome for cells within the liver consists of different enzymes that we use to break up and digest food. The proteome for red blood cells consists almost entirely of the hemoglobin that carries oxygen around our bodies. In all of these cases, the controlling factor for which part of the genome gets switched on is the environment in which the cell finds itself, and in particular the presence of near neighbors and the type of chemicals that they release or that are found on their surfaces.

From a physical scientist's point of view, that is as far as we can go. We know that the cells in an embryo can arrange themselves spontaneously into complex functioning patterns by following simple rules in response to signals from their nearest neighbors, and sometimes from more distant neighbors. We also know that the individual cells can also choose what functions to perform in the mature organism in response to other signals from their neighbors. What we do *not* know is whether this is all that "life" consists of or whether there is something more. No amount of extra scientific information is likely to resolve this issue. The argument is not a scientific one alone, although some scientists argue there is no point in believing anything that cannot be confirmed or refuted by scientific tests. To my mind, this is committing the same fallacy as Driesch did when he could only imagine one way that an intracellular machine might function and decided this was therefore the only way it *could* function. The fallacy in the first case leads to a hubristic materialism and in the latter case led to the equally hubristic philosophy of vitalism. The balanced point of view is surely to say that any belief (such as a mechanistic view of life) that can be tested against reality should be, but that this does not mean that beliefs that cannot be so tested are not true.

Driesch held his belief in vitalism to the end of his life, when he had given up science and had become president of the Psychical Society. He also maintained his scientific integrity and examined claimed psychic phenomena in as skeptical a manner as possible. As a philosopher, he believed that they existed; as a scientist, he would not accept claims that could be tested against reality until they had been so tested. Perhaps his true legacy is that he showed that the two forms of belief can exist within one person and that they need not be contradictory if the holder is honest about the nature of each.

8

Conclusion: Necessary Mysteries

The stories in this book have described some of the extraordinary beliefs that scientists now hold about the world and how it works and have shown how those beliefs came about. Many of them are *necessary mysteries* — bizarre, anti-commonsense beliefs that scientists have been forced to accept because they can't make sense of their observations without them. Few scientists would believe in a soul that has weight, but most now accept that weight itself is a mystery and believe that the mass of an object is probably conferred on it by a particle called a Higgs boson, which no one has ever seen.

In other cases, the mysteries concern things that no one will ever see. In their efforts to explain experimental results, scientists have come to have faith in the existence of a whole range of entities and effects that lie forever outside the range of our direct experience. We believe, for example, that heat, light, radio waves, motion, and even the work we do when we dig in the garden are all forms of *energy,* an entity that no one can ever experience directly, but only through its different manifestations. We also believe that electrically charged objects are surrounded by an invisible field of force, that light behaves as a wave even though it appears to travel as individual particles called photons, and that all materials are built up from atoms that are far too small ever to see or feel. All of these notions defied the common sense of their day, as did the idea that biological organisms might generate their own electricity for use in internal communications, and even in thought, or the notion that groups of molecules can "self-organize" themselves to form the structures of life. Scientists such as myself have been trained to accept these beliefs and to use them as tools to further our understanding of the world. They are not just the prerogative of the scientist, however. They are everyone's concern, and one of the aims of this book has been to make them accessible to people outside science by showing how some of these beliefs came about and why scientists continue to hold them.

The stories I have told of how scientists came to accept such notions are just the tip of the iceberg. Like Alice in Wonderland, present-day scientists can easily believe six impossible things before breakfast. A brief list of the more important anti-commonsense beliefs that scientists now hold is given in the appendix, which shows how scientists use them, and how they can influence our everyday lives in unexpected ways. That impact is not just concerned with practical issues. Our view of ourselves and our place in the universe has been profoundly affected by the discoveries and beliefs of science. This has made many people uncomfortable and even angry that science has taken away the essential mystery of life. Certainly science has removed some of the myths that resulted from ignorance, but I for one would not wish to go back to the days where disease, for example, was believed to be the result of witchcraft or "miasmas," rather than arising from concrete causes such as bacteria or viruses. Nor would I wish to live in an era where the authoritative pronouncements of a cleric or philosopher on how Nature *ought* to behave took precedence over direct observation of how Nature *actually* behaves.

This is not to say that experiment and observation can answer all of our questions about Nature (even though some scientists, with more hubris than sense, seem to think that it can). There is a whole host of questions that are too important to ignore but too difficult or impossible to answer from physical observation alone. These questions remain the province of philosophy and religion, and answers are only to be found with recourse to faith in a particular set of axioms or in a religious creed. Some scientists believe that recourse to faith rules such answers out of court, without recognizing that science, too, has its own brand of faith — the belief that an answer is likely to be true if the results it predicts are in accord with experimental observation. The more accurate the observation, and the more critical the questions the experimenter asks, the greater is the scientist's faith in the answer, but this does not mean that the answer is "right," or that the scientist has discovered "truth." Scientific theories can never really be proved to be true; we simply have faith in them to a greater or lesser extent depending on the difficulty and number of tests that they have passed.

The necessary mysteries that scientists now accept as a result of these tests lie alongside the other set of eternal mysteries that are the province of philosophy and religion. Their reality is overwhelmingly supported by experimental evidence and their existence, to my mind, constitutes very strong evidence for the existence of a world beyond our direct experience. What the nature of that world is, I have no idea, and I am not even sure what I mean by the concept. Perhaps it is the sort of world envisaged by some religions, or it may even be a physical world inhabited by souls that do have weight. Whatever it is, science needs it to understand the behavior of Nature in our own world, and in its attempts to do so science has provided us with a set of mysteries that are every bit as deep and interesting as those that have emerged from our attempts to answer the traditional great questions that will forever continue to challenge us.

Appendix

A Brief Catalogue of Necessary Mysteries

The following is a brief introduction to some of the major anti-commonsense beliefs that scientists now hold and use as tools in their everyday work. The stories behind some of them have been given in this book, and I have added others (in particular, the beliefs arising from relativity and quantum mechanics) whose stories will be more familiar but whose importance is undeniable.

1. Action at a Distance

What holds the universe together? Sir Isaac Newton's answer was *gravity,* a universal force that pulls all material objects, from stones to stars, toward each other. We have grown so used to the idea of gravity that it no longer seems mysterious, but it contains a mystery that stumped Newton, and which still puzzles scientists today — how can a force "reach out" across space to act on a distant object? Newton had no answer, which was why he famously said, "I make no hypotheses," meaning that he had no suggestions about how gravity worked and was only concerned with showing it *did* work. The idea of "action at a distance" that underlies it, however, still remains as one of science's necessary mysteries.

The law that Newton derived to describe how gravita-

tional forces between objects depend on their masses and on
the distance between them has many practical applications.
Scientists use it to predict the trajectories of balls, bullets,
rockets, and other free-flying objects that would travel in a
straight line at a constant speed except for the effects of gravity. It also has some other surprising consequences, as my father explained to me when I was a child. One of these is that
a person weighs less when the moon is overhead, because the
moon's gravity lifts the person slightly from the surface of the
earth.

Newton's Law of Gravity enters our daily lives in many
ways, from its effect on the swing of a pendulum to an athlete's ability to jump higher at high altitudes, where the pull
of Earth's gravity is less because the athlete is farther away
from the center of Earth. One important consequence of the
law arises if we fall off a ladder, where to fall twice as far
means that we will hit the ground twice as hard. So if a ladder starts to slip, it makes sense to scramble down a few steps
if possible so that the distance of the fall will be lessened and
the impact of the fall lowered. Better still, anchor the ladder
properly in the first place.

2. Forces

The mystery of "action at a distance" is not confined to gravitational forces. It applies equally to magnetic forces, the
forces between electrically charged objects, and the two other
principal forces of Nature — the "weak" and "strong" forces
that hold atoms and their nuclei together. How do all of these
forces come about? It seems that they are mediated by the

rapid exchange of "virtual" particles that can appear and disappear in the blink of an eye, but can live long enough in most cases to be identified and studied when atoms are smashed apart. Many such particles have now been discovered and sorted into families according to their possession of properties that include not only mass and electrical charge but also "color," "charm," "spin," and "strangeness." These words are labels (sometimes tongue-in-cheek) for underlying properties that appear to be necessary for the particle's function but whose real essence is outside the descriptive capacity of human metaphor as one of Nature's deepest necessary mysteries.

The physicists' picture of how forces work is called quantum chromodynamics. It underlies our efforts to understand how the universe was formed (not to mention how it might end) and is now a major tool in the search for practical nuclear power sources (regrettably for non-peaceful as well as for peaceful uses). We can get along perfectly well in our daily lives without any knowledge of quantum chromodynamics or even of the simple laws of force between electrical charges, but there is one particular circumstance where it is worth being aware of the consequences of those laws. It concerns the use of computers. Some computers have been ruined on very dry days when their users failed to realize how great an electrical charge could be built up on their own bodies by friction (with carpets, etc.). When they have touched their computer, they have delivered that charge to its sensitive innards. The answer is to discharge yourself by touching a grounded metal object such as a faucet before touching the computer, thus avoiding an unlikely, but potentially very expensive, problem.

3. Fields

The mystery of "action at a distance" between two objects led to many guesses as to how it might happen. One of the earliest was that each object was surrounded by a field, and that the effects occurred when the fields surrounding two objects overlapped and interacted with each other. We routinely talk of objects being surrounded by a gravitational field, an electrical field, or a magnetic field, for example. Their strength in different places is measured by the force that they exert on an appropriate test object — a piece of solid material in a gravitational field, a magnet in a magnetic field, or an electrically charged object in an electrical field. Some surprising "test objects" have also been used. The Bristol artist Richard Box, for example, has used isolated fluorescent light tubes to spectacularly demonstrate the presence of a strong electric field near high-voltage power lines (chapter 1).

Scientists still don't know what fields are. Some scientists believe that their reality or otherwise is a non-question, and the concept is only useful as a tool to describe mathematically how the force that one object can exert on another varies in space. In coffee room conversations, however, such scientists still tend to talk of fields as though they had an underlying reality. Whatever the case, fields still remain as one of science's most useful and necessary mysteries. The equations that described them are astonishingly accurate — to at least one part in a billion — and no exceptions have ever been found. These equations are routinely employed by scientists to design television aerials and remote communications systems (such as those used by mobile telephones), to control the movement of satellites, and to interpret the light from distant stars.

The concept of a field is so appealing that it has been

extended in popular culture to areas where its application is dubious, to say the least. On the day when I wrote the first draft of this chapter, I saw a notice outside a perfume shop that declared that "we are all surrounded by layers of invisible fields with different colors" (!). Perhaps people do produce some type of field as yet unknown to science (maybe even a "psychic" one), but the only reason to believe that this is the case would be if we could find, and preferably measure, some effect of that field on a distant person or other object, which is why scientists proposed the existence of such fields as the electric, magnetic, and gravitational in the first place. No claim for such an effect has *ever* been substantiated except for the following three cases.

4. Waves and the Ether

One way that an object might affect another at a distance would be if its movement was transmitted as a wave through some medium in which both were embedded. A speeding boat can affect the movement of another boat, for example, by creating a wave that travels through the intervening water to eventually cause the other one to bob up and down.

When Thomas Young proposed that light traveled as waves (chapter 3), this was the sort of picture that he had in mind. The medium in which the light wave traveled (called the ether) was seen as an intangible, jelly-like material that pervaded all of space, and the existence of this peculiar, but seemingly unobservable, material was regarded as yet another necessary mystery that accounted not only for the propagation of light waves but also for the action of fields, which were seen as distortions or "tensions" in the ether.

It soon became evident that light was only one of a family of waves that included radio waves, ultraviolet radiation, infrared radiation, and (later) X rays and microwaves. All of these differ only in the distance between successive waves and became known collectively as electromagnetic waves. In 1879 the brilliant Scottish physicist James Clerk Maxwell wrote down a set of equations describing how electromagnetic waves might travel through the ether via a myriad of counterrotating ether vortices that behaved like miniature gear wheels in space. Maxwell's equations showed that light must travel at a finite speed and that all other electromagnetic waves must travel at exactly the same speed — a prediction that has since been confirmed to astonishing precision by experiment. Maxwell's equations predicted other aspects of the behavior of electromagnetic waves that have been confirmed to a high degree of precision, and his equations are still used by physicists today, despite the fact that experiment has proved that the ether itself cannot exist! The mystery of the ether has hence been replaced by a deeper, but still necessary, mystery, which is that light (and all electromagnetic fields) behaves as though it is propagated through space as waves, but without the waves being "in" anything.

Maxwell's four simple equations are used by scientists to design telephone cables, interpret the results of brain scans, and understand the effects of sun spots. Hardly anyone outside of science has heard of them, but their influence on our daily lives is palpable, especially in the design of the dynamos that provide our electricity and the engines that convert it back into useful work.

5. Energy

Dynamos and engines convert one form of energy into another, but energy itself (chapter 1) remains one of the most important necessary mysteries. Who would have thought that light, heat, magnetism, electricity, X rays, radio waves, chemical transformation, and even the movement of physical objects would all have turned out to be different forms of the one thing — *energy*, an entity that no one has ever seen, touched, or experienced directly but which manifests itself here on earth in so many different ways?

The concept of energy is probably the most useful single tool in the scientist's repertoire. All of us use it on a routine basis, whether we are studying physics, chemistry, biology, geology, astronomy, or some arcane interdisciplinary combination. When I was measuring the forces between living cells, for example, I interpreted the results by using the fact that the force is no more than the rate of change of energy with distance, and it was the energy (i.e., the physical work) required to bring the cells together or pull them apart that I was really interested in.

Energy is one of the few scientific terms where popular terminology equates closely to scientific usage. It is a measure of the amount of physical work that can be done. Scientists have expanded the definition slightly by distinguishing between "total energy," "available energy," "free energy," and so on, but the real point is that all forms of energy can be converted into an equivalent amount of useful work under ideal conditions. When conditions are less than ideal, the actual amount of work obtained can be compared with the ideal amount to calculate the efficiency of an engine or a process. Plants, for example, convert light energy from the sun into the energy required for life processes with almost 100 percent efficiency, wasting hardly any energy as heat. If the process were any less efficient, the plant would overheat, shrivel up, and die. Other processes are less efficient. A typical car engine, for example, converts only 20 percent of the energy obtained from burning fuel into the mechanical energy required to move the car along. The rest is lost as heat.

When people in everyday conversation say, "I haven't got any energy," they mean what a scientist might mean, which is "I haven't got the capacity to do any physical work." The popular usage of the concept, though, like the use of the field concept, has expanded to encompass ideas that no reputable scientist would subscribe to. One of these is the idea of "psychic energy," so beloved of New Age thinkers, who seem to believe in it because, according to their philosophies, it *should* exist. This is perilously close to the sort of hubristic thinking that permitted the sixteenth-century Church to persecute Galileo because they believed that the universe *should* be perfect, and that it *should* be centered on Earth (chapter 2).

Actual observation revealed that Earth is not the center of the universe. It has never revealed the existence of such a thing as "psychic energy," and Bertrand Russell's tongue-in-cheek advice that "it is undesirable to believe a proposition when there is no ground whatever for supposing it true" seems in this case to be singularly apt.

6. Atoms and Molecules

Scientists now believe that all material substances, from seashells to stars, are made up of a relatively small number of different types of atoms. Some materials, called elements, are made up from atoms of a single kind (some 114 kinds are now known to exist). Most materials, however, are made of molecules, in which atoms are combined together in different ways. Some of the combinations are very simple. Water molecules, for example, contain just one oxygen atom and two hydrogen atoms (hence the formula H_2O). That formula, however, hides some fascinating complexities, including the fact that even this simple molecule has a shape:

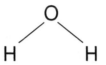

Figure A.1. The shape of the water molecule.
The lines represent "bonds" holding the hydrogen and oxygen atoms together, with the angle between the bonds being close to 105 degrees. Only with this angle can water molecules pack together in the open three-dimensional structure that is ice, with the openness of the structure ensuring that the ice floats on water rather than sinking. This feature of ice may not have been appreciated by the passengers on the *Titanic*, but without it it is doubtful whether life as we know it could exist on Earth.

Water molecules are tiny; less than a billionth of a meter across. Other molecules, which also have shapes, can be very much larger. A single molecule of human DNA, for example, exists in the form of the well-known double helix, contains some ten billion atoms, and would be around three centimeters long if unraveled.

Scientists are now so used to the idea of atoms and molecules that it seems hard to believe that their existence was doubted by some even up to a century ago. The original grounds for doubt were about what happened in the space between them, which seemed to be empty. "Nature abhors a vacuum" was the watchword, but it seemed that gases especially would be composed mostly of vacuum, with relatively few atoms or molecules whizzing back and forth and striking the walls of the container to create the effect of pressure. We now know this to be true — even when your car tires are pumped up to 30 psi or thereabouts, they still contain mostly empty space.

The idea that gases are mostly composed of empty space is so counterintuitive that it is a necessary mystery with which even Isaac Newton couldn't cope. He believed that atoms must have a form that would not leave empty spaces between them, and suggested that they were like springy balls of cotton wool packed in a jar. He was even able to show that Robert Boyle's famous relationship between gas pressure and gas volume followed directly from this assumption. Other evidence, however, showed that there was a great deal of empty space between atoms or molecules in a gas, and that Boyle's Law could equally well be derived from this assumption.

My mental world as a chemist is populated by atoms and molecules, whose shapes help me to understand the behavior of materials on a larger scale. My hair (of which there is not much left) is made up of molecules that are long and thin, while my skin is made from molecules that form sheets. In each case, the large-scale structure and properties reflect the structure and packing on the molecular scale, as they do in the cubic crystals of sugar, the slippery sheets of carbon atoms that form graphite, and the rigid three-dimensional structure built from carbon atoms that forms diamond. Even my difficulty in dislodging tomato ketchup from the bottle ultimately comes down to the way that the molecules in the ketchup are shaped so that they easily become entangled and the ketchup doesn't flow like a normal liquid (which is just as well once it gets onto the plate).

7. Momentum

Momentum is another word where common usage and scientific usage are remarkably close. When we speak of a football player's momentum carrying him over the line, for example, we mean that the football player is heavy enough, and moving fast enough, to make it difficult for players on the other team to stop him.

Scientists use the word *momentum* in much the same way, calculating it by multiplying a body's mass by its velocity. The product, however, has the truly amazing property that it is "conserved" — in other words, the momentum of a body can never change unless it is subjected to an external force (such as that imposed by a tackler). This fundamental law of physics, which forms the basis of Newton's Laws of Motion, has passed every experimental test, but the reason why (Mass × Velocity) should be such a fundamental quantity is another of Nature's necessary mysteries.

Early in the twentieth century, the German physicist Emmy Noether (one of the few women active in science at the time) discovered an even more stunning mystery that underlies the conservation of momentum. She showed that the principle follows directly from the fact that the laws of physics do not change under "translation of axes" — in other words, if a law applies when you are standing on the ground, the same law applies when you are flying above the ground at a constant speed in an airplane. This is why your sandwiches don't slide about on the tray or your drinks spill in your lap, even though the plane is traveling at more than five hundred miles per hour.

Emmy Noether went on to show that that the conservation of *angular momentum* similarly arises from the fact that the laws of physics do not change when the frame of reference is rotated. It is a sad reflection on the way in which science was (and is) a largely male preserve that Noether was never awarded a Nobel Prize for her perceptive insights.

The two momentum conservation laws (i.e., for linear and angular momentum), together with the law of the conservation of electrical charge and the law of the conservation of energy, form the bedrock on which physics rests. Physicists and engineers use these laws constantly as tools to analyze problems that involve the movement and collision of objects (chapter 2). In everyday life, the principle of the conservation of angular momentum finds particular use in the gyroscopic compass, where a spinning wheel maintains a constant direction that is independent of the earth's magnetic field or the pitching and rolling of a ship. It also governs the way that we ride bicycles or motorbikes, where we intuitively lean over as we turn a corner to compensate for the change in angular momentum.

The conservation of linear momentum determines how fast and in which direction the balls will go when we are playing pool. Sports commentators are apt to get the principle wrong, as with the cricket commentator I heard recently who said that a ball was "gaining momentum as it sped toward the boundary." Nothing gains momentum unless something is pushing it. When something does, beware. When push-starting a car, for example, be aware that it will keep moving at the same speed even after you have stopped pushing, so be ready to jump in if you are not already at the wheel!

8. Relativity

To understand Einstein's Special Theory of Relativity was once thought of as the pinnacle of intellectual achievement, yet it is now taught routinely to physics undergraduates and requires no special ability to understand the simple principle that underlies it. What it *does* require is a suspension of disbelief, because the necessary mystery on which the theory is based is so bizarre that it can never be understood in any conventional sense. Like many other necessary mysteries, it just has to be accepted, even though it defies the dictates of common sense.

Common sense says that, when two runners are racing each other, with the slower runner initially in front and the faster runner behind, then the faster runner will catch up to the slower runner at a speed that is the difference between their two actual speeds. That's all very well, unless the faster runner is replaced by the tip of a light beam. In this case, Einstein said, the speed of the leading runner doesn't enter into it. No matter how fast he or she runs, the light beam will still catch up at the same speed as if the runner were standing still.

The predictions of the Special Theory of Relativity have passed every experimental test. Some of these predictions are bizarre indeed, such as the notion that the faster you travel, the heavier you become (proved by experiments with electrons traveling at nearly the speed of light) and that a person who travels into space and then returns to Earth will be younger than if he or she had simply stayed on Earth (proved by experiments that compared the time registered by an atomic clock that was sent into space with that registered by an identical clock that remained on Earth). The theory also famously predicts that nothing can travel faster than the speed of light, which sets a limit on man's ambitions for space travel, among other things.

Not content with his Special Theory, Einstein later developed a General Theory of Relativity. You can test its basic

premise for yourself by taking a bathroom scale into an elevator and standing on it before the elevator starts. When the
elevator begins to move upward, the weight registered by the
scale will increase as a result of the elevator's acceleration. It
would also have increased, however, if your mass had increased (say by someone handing you a parcel to hold). Einstein argued that you could never tell whether acceleration or
a change in your mass has caused the measured weight to
change in a closed elevator and drew some startling and non-
obvious conclusions from this equivalence, such as that a beam
of light could be deflected by the gravitational field of a star
(now verified to considerable precision) and that gravity itself
is no more than a warp in the fabric of space, with time added
in just to make things interesting, so that "space" becomes
"space-time." His most astounding prediction, however, was
that mass is a form of energy, and that a small amount of matter may under suitable circumstances be turned into a large
amount of energy in the form of heat, light, and other forms
of radiation. Einstein was a pacifist and could not have known
that his prediction would form the basis for the atomic bomb.
It also forms the basis of modern attempts to produce energy
without pollution, although at the time of writing any technical success of these attempts seems some way off.

9. Quantum Mechanics

With his two theories of relativity, Einstein added considerably to our catalogue of necessary mysteries. Amazingly, he
did not win a Nobel Prize for either theory, but he did win
one, primarily for his elucidation of the mechanism of the
photoelectric effect, which provided an impetus for the de-

velopment of quantum mechanics—a theory in which he never truly believed!

Quantum mechanics, whose basic premise is that energy comes in lumps that cannot be further subdivided, provides a set of necessary mysteries that surpass in strangeness even those provided by the theories of relativity. For a start, it predicts that there is no such thing as a particle or a wave, and that all particles have wavelike characteristics, while all waves actually have particles associated with them. The two ways in which we can imagine that "action at a distance" might operate (by creating a disturbance in an intervening medium or by firing lumps of material across the intervening space) thus become merged into one. If you have difficulty understanding this, be comforted by the story of a promising postgraduate physics student who left physics for the easier profession of stockbroking. When his supervisor was asked why the student had given up physics, he replied sadly, "He tried to understand quantum mechanics."

Quantum mechanics simply can't be understood through metaphors derived from human experience. It has to be accepted as the way Nature actually behaves (however odd that behavior might appear to our eyes), because by accepting the premises of quantum mechanics we are able to understand many things we could not otherwise have understood, such as the composition of light emitted by the Sun, the emission of X rays and other radiation by decomposing atoms, and even the stability of atoms themselves.

Quantum mechanics is not only theoretically fascinating, it is also practically useful. It was used to predict the possibility of the laser, now used both as an industrial tool in engineering and as an essential element in CD players. It is routinely used by chemists to optimize the conditions for the

chemical reactions used to produce new drugs and other materials. It has even been used as the basis for a new form of
microscopy that permits us to visualize the shapes of those
molecules.

The technique is called scanning probe microscopy. It
relies on the fact that electrons with insufficient energy to

**Figure A.2. Researchers at IBM positioned forty-eight iron
atoms into a circular ring on the surface of a piece of copper in
order to corral some surface state electrons and force them into
quantum states of the circular structure, where they behave as
waves.**
The shapes of the peaks do not represent the actual shapes of the iron
atoms (these would be near-spherical), but they do show their positions.
In technical terms, the ripples in the center of the ring are the density dis-
tribution of a particular set of quantum states of the electrons that have
been "corraled." SOURCE: IBM.

jump across a very narrow gap may nevertheless cross the gap as a wave and reappear as a particle on the other side. This effect can be used to map out the contours of a conducting surface by passing a fine metal tip just above its surface and measuring the strength of the resulting electrical current. It has also provided a persuasive experimental picture of how particles and waves are two sides of the same coin, in an experiment where the tip was used to push iron atoms around on a copper surface until they formed a "corral," in which a set of waves was immediately seen to form in the center.

With this example, I come to the end of my brief catalogue of necessary mysteries. I have concentrated on examples from physics and chemistry, which underlie many of those to be found in biology and other scientific disciplines. I hope these examples will give the reader a taste of what it feels like to be inside the mind of a scientist and show that the universe is a more fascinating and surprising place than the human mind can, or ever will, envisage through metaphors based only on its own experience.

Notes

Preface

ix **explaining to a radio interviewer that the counterintuitive physical laws discovered by Galileo and Newton predicted that the wheels would stay on** The mad
exercise of riding a bicycle with the nuts removed from the
front wheel was a part of my BBC Radio 4 series *The Science
of DIY*. I was freewheeling along at a constant speed, and re-
lying on Newton's First Law of Motion (originally enunci-
ated by Galileo), which says that I did not need to apply a
horizontal force to keep moving. If I had applied a horizontal force, either by pedaling to speed up or braking to slow
down, I could well have displaced the wheel from the frame
and ended up in a tangled mess on the road. It's not an experiment that I would recommend anyone to try without a
husky "catcher" running alongside, and preferably not even
then.

xi **you don't have to be a genius to understand science —
it just needs persistence, and the wish to know** You
don't even have to be a genius to actually *do* science. Most science is done by people who are simply enthusiastic about understanding how the world works, and who use whatever
talents they happen to possess. A talent for mathematics is useful, but by no means essential. Other useful talents include
those for physical manipulation, categorization, and perception
of patterns. Their possessors could respectively have become
cabinetmakers, librarians, or artists, if their interests had tended

that way, but when they have entered science they have often made contributions that are every bit as significant as those of the well-publicized "geniuses" whose images unfortunately characterize the pursuit of science in the public mind.

1. Weighing the Soul

1 **a science call-in show on an Australian radio station** Karl Kruszelnicki (pronounced Crew-zel-inski) is the host. His program *Dr. Karl* on the rock radio station Triple-J attracts 15 percent of Sydney's population at eleven o'clock on a Thursday morning. Its popularity with his wide-ranging audience shows that the public thirst for understandable science is by no means confined to those whose usual choice is "serious" radio or TV.

1 **[Weight] tells us more about Nature than any other single measurement**   Archimedes started the trend some 2,200 years ago when he deduced that King Hiero's golden crown was adulterated with silver, because an equal weight of pure gold displaced less water from a full container than the crown did. This was the experimental idea that reportedly caused Archimedes to leap from his bath and run naked down the street shouting "Eureka!" As a teenager I did a lot of reading and thinking in the bath, and when I came across Archimedes' experiment in a book I decided to try it for myself, measuring the density of my own testicles by putting the bathroom scale on a chair and weighing myself, first in air and then with my testicles immersed in a sink full of water. The results were disappointingly inconclusive, but my failure at least led me to realize that accurate measurement was the key to experimental success.

The earliest instruments for measuring weight were the handheld balance scales used by craftsmen in pre-dynastic Egypt to assay gold and precious commodities. Indian street traders still use scales of a similar pattern, and many modern scientific scales use the same ordinary balance principle.

2 **the immaterial entity that we now call energy** The wheel came full circle with Einstein's revelation that matter itself is a form of energy.

2 **the jackal god Anubis using a set of balance scales to weigh the soul of a recently dead person against a feather** The Egyptian tomb artists usually represented the soul as a heart, thought to be the seat of the soul, but sometimes as a small person. Forty centuries later, medieval Christian art depicted the same process, with Anubis replaced by Saint Michael and the feather replaced by a toad. The toad represented the weight of the person's sins, although there are many variants on this theme. Examples of medieval pictures of Saint Michael weighing the soul are to be found in the oratory of St. Biagio in Perugia, Italy, (http://www.ukans.edu/ history/index/europe/ancient_rome/E/Gazetteer/Places/ Europe/Italy/Umbria/Perugia/Spello/Spello/churches/ S.Biagio/St.Michael.html) and in St. Botolph's Church, Slapton (near Peterborough, England) (http://www.painted church.org/slapweig.htm).

3 **Leonardo da Vinci was denounced as a sorcerer . . . for attempting to find the seat of the soul** Leonardo's accurate anatomical drawings of the dissected head, including the region supposed to house the soul, are held in the Royal Library at Windsor and are reproduced in *Leonardo: The Anatomy*, 3rd ed. (Florence: Giunti Gruppo Editoriale, 2000). Continuing the medical theme, my wife tells me that when she was a nurse in the 1960s it was still the custom in British hospitals to leave a body alone for an hour after death, before laying it out, to give the soul time to depart.

3 **a specific part of the brain . . . is associated with intense religious experience**   *New Scientist,* November 8, 1997, 7.

3 **I could not understand how a soul could touch and affect me if I could not touch and affect it**   This is the problem with Cartesian dualism, proposed in the seventeenth century by the French philosopher René Descartes, and described by the no-nonsense British philosopher Gilbert Ryle as "the ghost in the machine." According to the *Oxford Companion to Philosophy* (Oxford University Press, 1995), "The desire to avoid dualism has been the driving motive behind much contemporary work on the mind-body problem." Efforts to replace the ghost (e.g., by "psycho-physical parallelism," where consciousness is interpreted in terms of physical "brain states") have their own problems, however, and questions of the nature of consciousness and the related problem of the soul continue to preoccupy philosophers and to produce the occasional best-selling book.

3 **attempting to weigh the soul**   I am indebted to the psychologist and writer Dr. Susan Blackmore for directing me to MacDougall's experiments, which were the only verifiable efforts to weigh the human soul that I could find.

I am also indebted to the science fiction writer Brian Stableford for pointing out that there is a science-fiction novel called *The Weigher of Souls* by André Maurois, and that the scientist in Romain Gary's *The Gasp* attributes soul-weighing experiments to someone named Klaus working at the Württemburg Royal Institute in 1893, although neither Brian nor I have been able to find any further reference to this probably fictional character.

3 **a small municipal hospital in Haverhill, Massachusetts**   Now the Merrimack Valley Hospital.

3 **Duncan MacDougall . . . outlined his argument in a letter to his friend** MacDougall argued that the soul could only exist if it was a physical object, since otherwise he could not imagine what it might be. His argument is the inverse of the medieval Saint Anselm's famously flawed logical argument for the existence of God, which relies on the premise that if one can imagine something, then it must exist.

4 **scientists have repeatedly had to invoke the existence of things outside the range of our imagination just to make sense of the things that we can see, feel, and measure** This book tells the stories of some of the scientists who did the invoking. For a list of their counterintuitive notions and the relevance of their ideas to everyday life, see the appendix.

4 **the soul . . . must have material form** The Greek philosopher Democritus thought that the soul was composed of smooth, spherical "soul atoms," which flew apart at the moment of death, meeting with others to form different souls elsewhere. The philosopher Adam Morton tells me that he once had a correspondence with a man who claimed to be able to prove that the soul is composed of neutrinos, giving it a rest mass of zero and explaining why attempts to detect it *via* photography, absorption of electromagnetic radiation, or changes in temperature have been fruitless.

4 **MacDougall does not distinguish in his writings between weight and mass** *Mass,* measured in pounds or kilograms, is an intrinsic property of an object and is the same whether the object is on Earth, on the Moon, or in space. *Weight* is the downward force that the object exerts in a gravitational field on account of its mass. It is easier to lift a sixteen-pound bowling ball on the Moon than it is on Earth because the Moon's gravitational pull is less and so the downward force (i.e., the ball's weight) you need to overcome with an upward lift is less. If you were bowling the ball along an alley,

the force that you would need to use to accelerate the ball would be the same on the Moon as it is on Earth. Newton worked it all out with the simple relationship:

$$\text{Force} = \text{mass} \times \text{acceleration}$$

When the force referred to is the "weight" of the object, Newton's equation becomes:

$$\text{Weight} = \text{mass} \times (\text{acceleration due to gravity})$$

It is easy to confuse weight and mass because we tend to refer to both in units of mass. Old-time scientists sometimes got around the confusion by adding the word *weight* to the units of mass and referred to the downward force exerted by a mass of one pound or one kilogram respectively as a "pound-weight" or a "kilogram-weight." Better still is to refer to the weight in units of force. These days we measure that force in Newtons, units that are easily visualized, since one Newton is the force of Earth's gravity on an average-sized apple. More exactly, a mass of one kilogram exerts a downward force of 9.8 Newtons on Earth's surface.

5   **a standing jump . . . to a height of nearly six feet**   Pictures and details may be found in http://www.clavius.org/gravleap.html. For the scientific-minded, the height that a person can jump is given by:

$$\text{Height} =$$
$$(\text{vertical launch velocity})^2 / (\text{twice the acceleration due to gravity})$$

This formula means that a person can jump six times as high on the Moon as he or she can on Earth, since the Moon's gravitational pull is one-sixth of that on Earth. Even so, Neil Armstrong must have been pretty fit — I'm not sure I could do a standing jump of even one foot on Earth when dressed in a space suit.

6 **The winning entries [to the "Higgs boson" competition]** The entry by Mary and Ian Butterworth (Imperial College, London) and Doris and Vigdor Teplitz (Southern Methodist University, Dallas, Texas) read as follows (*Physics World* 6, no 9 [1995]):

The Higgs boson is a hypothesized particle which, if it exists, would give the mechanism by which particles acquire mass.

Matter is made of molecules; molecules of atoms; atoms of a cloud of electrons about one-hundred-millionth of a centimeter and a nucleus about one-hundred-thousandth the size of the electron cloud. The nucleus is made of protons and neutrons. Each proton (or neutron) has about two thousand times the mass of an electron. We know a good deal about why the nucleus is so small. We do not know, however, how the particles get their masses. Why are the masses what they are? Why are the ratios of masses what they are? We can't be said to understand the constituents of matter if we don't have a satisfactory answer to this question.

Peter Higgs has a model in which particle masses arise in a beautiful, but complex, progression. He starts with a particle that has only mass, and no other characteristics, such as charge, that distinguish particles from empty space. We can call his particle H. H interacts with other particles; for example, if H is near an electron, there is a force between the two. H is of a class of particles called "bosons." We first attempt a more precise, but non-mathematical statement of the point of the model; then we give explanatory pictures.

In the mathematics of quantum mechanics describing creation and annihilation of elementary particles, as observed at accelerators, particles at particular points arise from "fields" spread over space and time. Higgs found that parameters in the equations for the field associated with the particle H can be chosen in such a way that the lowest energy state of that field (empty space) is one with the field not zero. It is surprising that the field is not zero in empty space, but the result, not an obvious one, is: all particles that can interact with H gain mass from the interaction.

Thus mathematics links the existence of H to a contribution to the mass of all particles with which H interacts. A picture that corresponds to the mathematics is of the lowest energy state, "empty" space, having a crown of H particles with no energy of their own. Other particles get their masses by interacting with this collection of zero-energy H particles. The mass (or inertia or resistance to change in motion) of a particle comes from its being "grabbed at" by Higgs particles when we try and move it.

If particles do get their masses from interacting with the empty space Higgs field, then the Higgs particle must exist; but we can't be certain without finding the Higgs. We have other hints about the Higgs; for example, if it exists, it plays a role in "unifying" different forces. However, we believe that nature could contrive to get the results that would flow from the Higgs in other ways. In fact, proving the Higgs particle does not exist would be scientifically every bit as valuable as proving it does.

These questions, the mechanisms by which particles get their masses, and the relationship among different forces of nature, are major ones and so basic to having an understanding of the constituents of matter and the forces among them that it is hard to see how we can make significant progress in our understanding of the stuff of which the earth is made without answering them.

Science, in other words, has not yet run out of major questions to answer!

6 **Fairbanks Standard platform scales** The platform scale was invented by Thaddeus Fairbanks in 1830, and scales of a similar pattern are still manufactured by the Fairbanks Company in the United States today.

8 **Have I discovered [the] soul substance . . . ?** The idea that there *was* an actual "soul substance" seems first to have been proposed by the physiologist Rudolph Wagner at the Göttingen Congress of Physiologists in 1854.

11 **he wrote an account of his experiments** MacDougall's account was grandly entitled "Hypothesis Concerning Soul Substance Together with Experimental Evidence of the Existence of Such Substance." He published it in two places. One was *American Medicine* for April 1907. The other was the *Journal of the American Society for Psychical Research* 1, no. 5, (1907): 237–44. The editor of the latter journal commented, "It was not his [MacDougall's] intention that his experiments should obtain public notice at present, but an unauthenticated publication of his attempts, with the usual distortion that everything gets in the papers, has resulted in this prompt effort to correct the misrepresentation."

13 **That is the way science works; at least, it is the way that it is supposed to work** The biggest problem with secrecy in science, whether the secrecy is demanded by industrial or military sponsors, is not that the results are unavailable to others but that they are unavailable to be *criticized* by others. The inevitable result is that much of military and industrial science is, well, bullshit.

13 **"Extraordinary claims require extraordinary proof"** This powerful criterion is usually attributed to the astronomer Carl Sagan, who used it in his book *Cosmos*. It was first used two centuries earlier, though, by the skeptical Scottish philosopher David Hume, whose essay "On Miracles" should be required reading for all modern believers in UFOs, astrology,

spoon-bending, and even the claims of some scientists. Hume was concerned with the value of testimony as evidence and concludes:

> No testimony is sufficient to establish a miracle, unless the testimony be of such a kind, that its falsehood would be more miraculous, than the fact, which it endeavours to establish.

Modern research on testimony indicates that no two witnesses remember an event alike, and that even the most convincing testimony must be taken with a grain of salt, let alone testimony that has been passed down from generation to generation.

13   **the supposed discovery of a new form of water**   This was "polywater," formed by the condensation of water vapor in narrow silica tubes and supposedly having different physical properties from normal water. My colleagues still pull my leg about a paper that I published in my youth (it was my third ever scientific paper), which showed that a substance with similar spectral properties could be formed by the condensation of water on flat surfaces that did not contain silica, and which I unwisely identified with "polywater." That said, the properties of water near silica and other surfaces still pose some of the most baffling unsolved problems in surface science.

13   **Similar experiments were performed in the 1930s on some unfortunate mice**   Twining reported his experiments with mice in H. Carrington, *Laboratory Investigations into Psychic Phenomena* (London: Rider and Co., 1939). Susan Blackmore provides an excellent summary in her book *Beyond the Body* (London: Heinemann, 1982).

14   **the first member of the Aero Club of California to build a monoplane that flew**   With his partner, Colonel Warren Eaton, in 1910.

14 *The Physical Theory of the Soul*   Published by the Press of
the Pacific Veteran, Westgate, California, in 1915.

15 **The fluid was even given a name — *caloric***   The name
was suggested by Guyton de Morveau in 1787 and adopted
by Lavoisier in his book, which was published in 1789.

16 **Count Rumford . . . started life as plain Benjamin
Thompson in America**   I cannot resist repeating the story
that I told in *How to Dunk a Doughnut* about Thompson's in-
volvement in the American War of Independence, where he
was in charge of British troops who used gravestones from a
local cemetery to build a bread oven. They distributed the
excess loaves to the local community, who were dismayed to
find the names of their dead relatives baked backward into
the crusts.

16 **Rumford . . . designed an experiment [to weigh heat]**
Benjamin Count of Rumford, F. R. S. M. R. I. A. &c., "An
Inquiry Concerning the Weight Ascribed to Heat," *Philosophical Transactions of the Royal Society of London* 89 (1799):
79–194.

16 **heat is expected to leave a liquid when it freezes**   and
vice versa — it takes the same amount of heat to turn solid
ice into liquid water as has to be removed from the water to
turn it back into solid ice. The heat involved is called latent
heat, a term coined by the Scottish doctor Joseph Black and
a concept used to great practical effect by Black's pupil James
Watt in his invention of the steam engine.

17 **keep my gin and vodka in the freezer**   London dry gin
contains approximately 47 percent alcohol by volume, and
will only freeze below -33°C. A typical domestic freezer is set
at -18°C, which gives a reasonable safety margin. Don't try
this with beer, which freezes at around -2°C, because the water expands as it freezes, cracking the glass. This happened to

me once in Australia when I forgot that I had put a bottle in the freezer to cool down quickly, but such was my thirst that I peeled the shards of glass off the outside and ate what was to all intents and purposes a beer ice-block.

The notion of "freezing point" is often misunderstood. It is the unique temperature at which liquid and solid can co-exist. Below the freezing point, all becomes solid, which rather makes nonsense of a London fireman's claim after rescuing a child from icy waters that "the water was freezing, and it was probably even colder than that."

18 **replaced the spirit of wine in one bottle with mercury**
Rumford's idea was that "the quantity of heat lost by the water must have been very considerably greater than that lost by the mercury; the specific quantities of latent heat in water and in mercury, having been determined to be to each other as 1,000 to 33; but this difference in the quantities of heat lost produced no sensible difference on the weights of the fluids in question."

In other words, it takes a lot more heat to freeze water and lower its temperature by a few degrees than it does to lower the temperature of mercury by the same amount, but this difference did not affect the result.

19 **belief in science or religion is a complex business**
Some religious beliefs, such as the Bible story of the creation or Christian belief in the soul, have some overlap with the beliefs of science. Where the two overlap, controversy follows, as MacDougall (and Galileo and Darwin before him) found to their cost, and as many modern-day scientists still do. It is not the purpose of this book to enter into these controversies, on which more than enough has already been written, but to point out that science itself is based on beliefs that have evolved as they have been tested against reality.

It might surprise some readers that a scientist should be writing about "belief." Isn't science just a matter of hard, soulless, irrefutable fact? Isn't it, in fact, just "applied common sense," as the great Victorian biologist T. H. Huxley put it? The simple answer is "No," and the stories of the scientists in this book show how very wrong Huxley's definition was. Throughout history, many scientists have tried to use common sense to understand how the world works — and failed. The scientists in this book had the courage to believe in the existence of things outside our experience as the only way to make sense of that experience. Their ideas have given us access to a whole world outside our experience, populated by entities and that we can never touch, see, or feel but whose existence we must accept if we are to make sense of the things that we can touch, see, and feel. By revealing the existence of that world, science in this sense provides the strongest evidence we have that a world beyond our sense experience exists, even if it will never be able to tell us whether the soul is a part of that world.

21 **the amount of light that I can get from the laser will be the same as that which I originally shone onto the photocell (with an allowance for loss of some energy as heat during the two processes)** Further exploration of this point is beyond the scope of this book, and would involve thermodynamics, a subject on which I once wrote a whole textbook chapter. When I came to reread the chapter a few years later, I couldn't understand a word of it.

21 **light can be used to spin a small fan called a Crookes' radiometer** The "fan" consists of four lightweight metal vanes, polished on one side and blackened on the other. It is contained within an evacuated glass bulb and spins when exposed to light. Crookes' radiometers are readily available commercially and are usually accompanied by a leaflet that gives totally the wrong explanation for the spinning — that the impact of photons of light bouncing off the shiny side causes the fan to spin. If the writers of these explanations had

taken the time to look, they would have found that the fan spins in the wrong direction for this "explanation" to hold. The truth is that light absorbed by the blackened side makes this side hotter than the shiny side, so that air molecules near the edges of the blades on the blackened sides are heated more by radiation and hence are moving faster when they strike the blade than are air molecules near the shiny sides. The resultant difference in pressure causes the fan to spin.

21   **movement is also a form of energy, called kinetic energy**   The heat that I generated in rubbing my hands together was no more than the microscopic increase in kinetic energy of the myriads of molecules in the skin of my hands.

21   **store and transport it [energy] in one form and then convert it and use it in another form**   For example, we store energy in our cars in chemical form as gas, which we then burn to produce heat, which we then use to drive an engine and move a car.

2. Making a Move

23   **Aristotle's ancient ideas**   In his book hubristically titled *Physics,* published in 350 B.C., Aristotle promulgated rules on motion that held sway for two thousand years. One of the fundamental propositions of Aristotelian philosophy is that there is no effect without a cause. Applied to moving bodies, this proposition dictates that there is no motion without a force. Speed, according to Aristotle, is proportional to force. This notion is not at all unreasonable if one takes as one's defining case of motion, say, an ox pulling a cart: the cart only moves if the ox pulls, and when the ox stops pulling the cart stops. What Aristotle didn't realize was that the cart only stops because of friction — without friction, it would keep on rolling.

23 **the trolley kept on rolling**   My own record is not so un-
blemished that I can afford to reveal either the name of my
highly placed source for this story about a group of scientists
who should have known better or the name of the prestigious university at which they worked.

23 **a junior lecturer at the University of Pisa**   Pisa was
Galileo's hometown. When he was born there in 1564, the
famous Leaning Tower had already been in existence for four
hundred years, and was leaning a perilous four meters to the
south.

24 **Pope Urban VIII,**   then known as Maffeo Barberini, had
been a fellow student of Galileo's at the University of Pisa,
with Galileo studying medicine and Barberini studying law.
For an informative and entertaining account of the conflict
between them in their later years, see Hal Hellman, *Great
Feuds in Science* (New York: John Wiley, 1998). A more detailed, but equally interesting account is given by Maurice
Finocchiaro in *The Galileo Affair* (Berkeley: University of
California Press, 1989).

24 ***Dialogues Concerning Two New Sciences*** *(Discorsi e di-
mostrazioni matematiche intorno a due nuove scienze attenenti alla
meccanica)* is available in English translation at http://www.
phys.virginia.edu/classes/109N/tns_draft/index.html.

24 **Italian History of Science Museum . . . Room IV**   For
a virtual visit to this treasury of Galileo's inventions and
discoveries, see http://galileo.imss.firenze.it/museo/4/index.
html.

25 **We see Galileo now as a rebel**   His father certainly did.
He wanted Galileo to be educated for college and to become
a physician, but when the young Galileo was sent to a
monastery school he became so enamored of the life there

that he decided to become a monk at the age of thirteen. He was eventually persuaded to enter the medical school at Pisa, but he skipped most of his medical lectures and spent his time studying mathematics. His tutor eventually persuaded Galileo's father to let Galileo leave Pisa without a degree and take up a career teaching mathematics.

25 **several years as a private mathematics tutor** Galileo did much of his teaching at Vallombrosa, then as now a monastic center that provided a refuge for escape and contemplation. Similar centers for scientists still exist, but the competition to enjoy the opportunities that they offer can be fierce. I was very excited when I once managed to wangle three weeks at the Aspen Center for Physics in Colorado but more than a little intimidated when I arrived to find that roughly a third of the forty or so scientists there had Nobel Prizes. They were all very nice to me, though, and I learned a huge amount from talking with them.

25 *La Bilancetta (The Little Balance)* This described an improved method for weighing precious metals in air and then in water, and hence determining the gold content. The part of the arm on which the counterweight was hung was wrapped with metal wire. The amount by which the counterweight had to be moved after immersing the object in water could then be determined very accurately by counting the number of turns of the wire, and the proportion of gold to silver in the object could be read off directly.

26 **dimensions and location of Hell** The full text of Galileo's lectures, including his recapitulation of Dante's journey, can be found in Galileo, *Circa la figura, sito e grandezza dell' Inferno di Dante*. Here is his description of Hell in his own words:

[A]s far as the shape is concerned, I say that this is the form of a concave surface called conic, the peak of which is at the center of the world, with the base towards the surface of the earth. But then, let us abbreviate and simplify the reasoning. Linking shape, position, and size, let us imagine a straight line stretching from the center of the earth (which is still the center of gravity and of the universe) as far as Jerusalem, and an arc which extends from Jerusalem above the surface of the water and earth for a twelfth of its major circumference. Such an arc will terminate with one of its extremities in Jerusalem. If, from the other end, another straight line were to be drawn as far as the center of the world, we would have the sector of a circle, contained by the two lines coming from the center and from the said arc. Let us then imagine that, the line joining Jerusalem and the center being immobile, the arc and the other line are set in motion, and that with such a movement it slices the earth, and keeps moving until it returns to its starting point. A part similar to a cone would be cut out of the earth, and if we imagine this to be removed, there would be left, in its place, a hole in the form of a cone; and this is Hell.

Galileo further deduced, by extrapolating from Dante's line that "[the giant Nimrod's] face was about as long / And just as wide as St. Peter's cone in Rome," that Lucifer himself was two thousand arm-lengths long. Interestingly, the "mouth" of the Hell that Galileo calculated, and that Lucifer was said to inhabit, passed through Naples. The old saying "See Naples and die" might have more meaning than we realize!

Galileo analyzed Dante's poetry, but others have attempted to use mathematics to analyze the Bible itself. A famous example is the Irish bishop Usher, who in 1645 painstakingly worked out the lifetimes of biblical characters, and concluded from adding them up that the earth must have been created on October 26, 4004 B.C. When I was given my first Bible at Sunday school, this figure was quoted as an established fact in a note at the head of the Book of Genesis.

On a rather more frivolous note, an anonymous scientist recently used mathematics to "prove" that Heaven is hotter than Hell. His argument was that Hell contains a lake of liquid sulfur (Revelations 21:8) and therefore must have a temperature below the boiling point of sulfur (444.6°C). Heaven, though, receives forty-nine times as much radiation from the sun as does the earth (Isaiah 30:26), which, by the laws of radiation, means that Heaven must have a temperature of 525°C. *Ergo,* Heaven is hotter than Hell. See *A Random Walk in Science* (London: Institute of Physics, 1973).

28 **the large piece would rise to the surface of the water a hundred times more swiftly**   This sentence gives a clue as to how Galileo thought about the problem of falling bodies, which was to begin with something that he understood thoroughly (flotation) and to apply his knowledge to the new problem. Opportunities abound for scientists to think in this way. My first boss was an expert in electronics, but his job was to predict the flow of heat in different engineering situations. Luckily for him, the equations for heat flow had exactly the same form as the equations for the flow of electricity — only the symbols meant different things. Given a problem in heat flow, he would build the equivalent electrical circuit, measure the appropriate voltages and currents, and convert the answer to units of heat. The process is called simulation, and the electrical circuit was called an analog computer. It is amusing to note that analog computers themselves are now simulated by today's digital computers.

Most scientists develop tricks where they apply a thorough knowledge of one area to problems in other areas. The ultimate exponent of this technique was the Italian Enrico Fermi, developer of the atomic pile and modern inheritor of Galileo's mantle. Fermi succeeded in reducing the whole of physics to just seven equations, and could recast any problem in the form of one of these equations to rapidly produce the answer. (Life can be so simple for some people!)

Fermi was a genius. So was Galileo, but even geniuses can get off on the wrong foot. This is what happened to Galileo when he attempted to apply the principles of flotation to the problem of falling bodies, because he initially substituted for Aristotle's error an error of his own, namely that the rate of fall should not depend on the weight, but should depend on the density.

28 **two bodies whose natural speeds are different** It was an Aristotelian idea that bodies had "natural places" that they strived to reach and "natural speeds" at which they took the journey. Weighty bodies, according to Aristotle, strove to fall toward the center of Earth, and heavier bodies did this at a higher speed than lighter bodies.

29 **The method of "proof by contradiction"** This was used by Euclid to establish that there is no such thing as a largest prime number, because if there were, we could multiply it and all the prime numbers below it and add one to the result to produce a still larger prime number (i.e., not divisible by any of the prime numbers used in the multiplication), which contradicts our original assumption. See G. H. Hardy, *A Mathematician's Apology* (Cambridge: Cambridge University Press, 1940), reprinted with an introduction by C. P. Snow, for a fuller explanation, and a wonderful description of what it feels like to be a mathematician.

29 **But I, . . . who have made the test** Galileo is speaking here in the guise of Salviati in *Dialogues Concerning Two New Sciences*. He was not actually the first to perform such tests. Guiseppe Molti, whose chair he was later to take over at Padua, had already shown experimentally that Aristotle was wrong, and that bodies of the same material but different weight arrive at earth at the same time when dropped from

the same height. Molti had also shown that bodies of the same volume but different material fall to earth at the same rate, which contradicted Galileo's first predictions.

30 **The Campanile di San Marco in Venice is a candidate** Another possibility is the Torre degli Asinelli in Bologna, at a height of nearly 100 meters.

30 **a group of students at Rice University replicated Galileo's original experiment**   The repetition of Galileo's "falling ball" experiment is described in http://es.rice.edu/ ES/humsoc/Galileo/Things/on_motion.html. It was performed as part of the "Galileo Project," a project of Rice University whose aim is to provide detailed information on the life and work of Galileo Galilei (1564–1642) and the science of his time.

31 **These are bottles fit only to pee into**   Quoted in http:// inventors.about.com/library/inventors/blgalileo.htm.   Modern scientists continue to be scathing about their colleagues, although the criticisms are not always quite so direct. A colleague of mine once attended a lecture where the Nobel Prize winner Salvatore Luria was sitting in the front row. After a few minutes Luria put his hand up and asked the lecturer, "Do you really believe that?" "Yes, Professor Luria," answered the lecturer, and kept on going. After a few minutes Luria quietly left the auditorium, but as he reached the back door he stuck his hand into the projector beam and gave the thumbs-down sign.

31 *De Motu*   Galileo made incredible progress in understanding the motion of falling bodies when one considers that he did not have the concept of gravitational attraction to help him. This concept was only introduced by Isaac Newton, who was born in the year that Galileo died.

31 **rolling them down an inclined plane**   The "law of falling bodies" that Galileo finally elucidated with this technique was one of his most important contributions to science. The way that he elucidated it is very typical of how science usually works, which is not by brilliant "breakthroughs" alone, but by patient testing and refining of an initial idea that might be proved right but whose value is often simply to provoke the experimental testing that eventually leads to the right idea.

32 **if you fall off a high ladder, you will have fallen four times as far after two seconds as you had after one second**   With the aid of Newton's Laws, you can also calculate that you will be traveling only twice as fast after two seconds as you were after one second, even though you will have fallen four times as far. This is no comfort, however, because your kinetic energy depends on the square of your speed, so if you hit the ground after two seconds the impact will be four times as great as if you struck it after one second.

32 **Sagredo**   At one of his parties he also introduced Galileo to the Murano glassmaker who would later craft optical lenses for his telescopes.

32 **Marina Gamba**   Living in Galileo's house in Padua, she bore him three children. His two daughters, Virginia and Livia, were both put in convents where they became, respectively, Sister Maria Celeste (subject of *Galileo's Daughter* by Dava Sobel) and Sister Arcangela. In 1610, Galileo moved from Padua to Florence, where he took a position at the court of the Medici family. He left his son, Vincenzio, with Marina Gamba in Padua. In 1613, Marina married Giovanni Bartoluzzi, and Vincenzio joined his father in Florence.

32 **It was at Salviati's . . . villa**   This productive combination of science and social activity was not unusual then, and it is not unusual today. Scientists thrive on company, and the

coffee room, the pub, and the party play just as important a part in our lives as the laboratory or the library. When I was working as a scientist in Australia a group of us took to organizing seminars at eight o'clock on a Friday morning on Sydney's Coogee Beach. We would plunge into the surf and then return to fuel our minds with champagne and croissants, discuss ideas, and draw equations in the sand. Many good ideas emerged in this relaxed environment, and on one occasion an idea proved so exciting that I did not notice that I was still covered in sand when I rushed back to the lab to try it. One of our group later wrote about some of our beach exploits in a Sydney newspaper.

33 **the telescope . . . was a Dutch invention** The honor seems to belong to Hans Lipperhey, a Dutch spectacle maker who came up with the idea of putting lenses at either end of a tube so as to combine their magnifying power. He applied for a patent to his local Zeeland government office, and was sent to the central government office with a letter asking them to be of help to the bearer "who claims to have a certain device by means of which all things at a very great distance can be seen as if they were nearby, by looking through glasses which he claims to be a new invention." The military potential was obvious, and the patent was denied because it was felt that the device could not then be kept a secret, but Lipperhey was paid handsomely to make more telescopes.

Shortly after Lipperhey's application, the States General were also petitioned by Jacob Metius of Alkmaar, a city in the north of the Netherlands; he also claimed to be the inventor. The claim of yet a third person, Sacharias Janssen, also a spectacle maker in Middelburg, emerged several decades later. The surviving records are not sufficient to decide who was the actual (or as it was put in the seventeenth century, the first) inventor of the telescope. All we can say is that Lipperhey's patent application is the earliest record of an actually existing telescope.

33 **too many people thinking along parallel lines**   One of the problems with prizes like the Nobel Prize is that it is unusual for just one scientist to have made a "discovery." Most discoveries require a series of steps, and the prizewinner is often the one who happened to make the final step, or who pulled all the other steps together.

35 **James Thurber . . . tried to see plant cells down a microscope**   "University Days" in *My Life and Hard Times*, repr. (New York: Perennial, 1999).

37 *Dialogue Concerning the Two Chief World Systems* **[Ptolemaic & Copernican]** *(Dialogo sopra i due massimi sistemi del mondo, tolemaico e copernicano)*. The full text is available at http://webexhibits.org/calendars/year-text-Galileo.html.

37 **the pope was furious**   Even after Galileo's death, the pope blocked plans for a public funeral and a monument at the church of Santa Croce in Florence. Instead, Galileo's remains were quietly hidden away in the basement of the church bell tower, where they stayed for almost a century.

37 *Index Librorum Prohibitorum*   The first edition of the index of books "that can be dangerous to faith and morals" was published by Pope Paul IV in 1557. The 32nd edition was published in 1948 and contained 4,000 titles, including Victor Hugo's *Les Misérables,* Gibbon's *Decline and Fall of the Roman Empire,* Goldsmith's *Abridged History of England,* Immanuel Kant's *Critique of Pure Reason,* and all of the works of Jean-Paul Sartre. The *Index* was finally suppressed in 1966.

37 **returned to the basic studies**   The range of questions that Galileo addresses in *Two New Sciences* is truly amazing. He suggests a method for weighing air, and provides a wonderful trick method for measuring the length of a vertical piece of string "whose upper end is attached at any height whatever even if this end were invisible." He even wanders into my

own area of specialty by considering how water droplets sit up individually on plant leaves and don't spread out and run together, as they would on a clean glass surface. This is one of the few questions for which he does not have an answer, although this is hardly surprising, as it took the development of a whole new scientific discipline (thermodynamics) to understand this complex phenomenon.

Above all, *Two New Sciences* laid out a program for much of subsequent physics. It even presages the invention of calculus by trying to work out whether the rolling of a polygon became closer and closer to the rolling of a circle as the number of sides became larger and larger. Galileo also came up with a beautiful argument that light must travel with a finite speed:

> [We] see the lightning flash between clouds eight or ten miles distant from us. We see the beginning of this light — I might say its head and source — located at a particular place among the clouds; but it immediately spreads to the surrounding ones, which seems to be an argument that at least some time is required for propagation; for if the illumination were instantaneous and not gradual, we should not be able to distinguish its origin — its center, so to speak — from its outlying portions.

Not content with theoretical argument, he dreamed up and tried an experiment where two observers alternately flashed lights at each other from different distances to try to measure the speed of light. Light travels too fast for his method to work, but the speed was eventually measured by a remarkably similar method using one observer, a mirror, and a rapidly rotating toothed wheel to alternately block and admit the light.

He was even up to party tricks:

> I was able to deceive some friends to whom I had boasted that I could make up a ball of wax that would be in equilibrium in water. In the bottom of a vessel I

placed some salt water and upon this some fresh water; then I showed them that the ball stopped in the
middle of the water, and that, when pushed to the
bottom or lifted to the top, [it] would not remain in
either of these places but would return to the middle.

This wasn't just a party trick. The idea proved useful to
physicians who were trying to measure slight changes in density of different bodily fluids, and the method — called a
"density gradient" — is still used in modern molecular biology to separate different types of living cell with slightly different densities.

Two New Sciences shows the scientist at work, warts and
all. To me it revealed that Galileo went through the same
struggles to work out the basic rules of motion that I went
through as a student when I was trying to understand those
rules, and that many people outside science go through when
trying to understand how the world works for themselves. The
ideas that I had been taught as sterile fact came alive when I
read how Galileo thought about them for the first time.

42 **Churchill stated categorically** See Brian Marriner,
Forensic Clues to Murder (London: Arrow Books, 1991), 135.

3. A Salute to Newton

45 **the Galileo Space Probe** For full information about the
Galileo mission to Jupiter, see http://galileo.jpl.nasa.gov/.

45 **A series of dark lines will mysteriously appear in the
gap, apparently suspended in space** My colleague Peter
Mason showed me this experiment when we were traveling
together in a train crowded with well-lubricated Australian
football fans. One of them asked irately what Peter was doing
holding two fingers up to him (the Australian equivalent of a
one-fingered obscene gesture in America), and Peter hastily

explained that he was demonstrating a scientific experiment. The aggressor tried the experiment for himself, and was very impressed by the appearance of the lines. He was delighted to have successfully performed a real scientific experiment, and went around showing it to his mates. When we left the train, our car was full of people holding up two fingers.

Peter Mason was the Foundation Professor of Physics at Macquarie University in Sydney, and was one of the bravest men that I have ever met. He was a brilliant expositor of popular science and continued to give regular radio broadcasts on the subject even while he was dying of a brain tumor, describing the diminution in his own speech as the tumor developed and speculating on its physical causes. When he gave his last broadcast, his vocabulary had dropped to fewer than one hundred words. This chapter is dedicated to his memory.

46 **the ripples behind two swans passing on a lake** Thomas Young's original inspiration for thinking of light as waves may actually have been the observation of interference patterns between the ripples in the wakes of two swans swimming past each other on the lake at Emmanuel College, Cambridge, where he was a medical student. J. D. Mollon, "The Origins of the Concept of Interference," *Philosophical Transactions of the Royal Society of London* A360 (2002): 807–19.

46 **Thomas Young . . . was terribly serious-minded**   If Young had a sense of fun, it is certainly not to be found in his biography, *Life of Thomas Young, M.D., F.R.S., &c.,* by George Peacock (London: John Murray, 1855), a mid-nineteenth-century panegyric that reveals his biographer had as little sense of humor as Young himself.

47 **Francis Davenport sent a record of [the tides]**   This occurred at a time when Edmund Halley was busy finishing his catalogue of the southern skies (*Catalogus stellarum australium,* 1678), which may also explain the delay in publication. Shortly after he had published the letter and his explanation

("An Account of the Course of the Tides at Tonqueen in a Letter from Mr. Francis Davenport July 15, 1678, with a Theory of Them, at the Barr of Tonqueen, by the Learned Edmund Halley Fellow of the Royal Society," *Philosophical Transactions of the Royal Society* 14 [1684]: 677–88), Halley was forced to take up a menial position at the society as assistant to the secretaries of the society following the mysterious murder of his father, who had provided his financial support.

Halley's "complicated equation" was based on Davenport's observation that the tide was highest when the Moon was at its maximum declination. It said that "the *increase* of the *waters* should always be proportionate to the *Versed signes* of the doubled distances of the *Moon* from the *Equinoctial* points."

48 **[Newton's]** *Principia* Halley wrote the foreword for the *Principia* (published in 1687) and must have been galled to find that all his work on an equation for the tides was essentially meaningless.

48 **[Newton's]** *Opticks* is much more readable than the *Principia* and was written in English rather than Latin. The version that I use and quote from is the Dover edition prepared by Bernard Cohen, published in 1979.

49 **Newton had poor eyesight** See Peter Mason, *The Light Fantastic* (Sydney: Penguin Australia, 1981).

52 **If Newton had been able to factor in the Earth–Sun distance** The distance was finally worked out in 1769 by a method suggested by Halley, which was to time the transit of the planet Venus across the face of the Sun as it passed between the Sun and Earth. Captain James Cook stopped off at Tahiti to do the experiment on June 3, 1769, before going on to discover Australia.

52 **Brougham set himself to confirm Newton's ideas experimentally** Brougham claimed complete objectivity in

the interpretation of his experiments, saying "He who obstinately adheres to any set of opinions, may bring himself at last to believe that the fresh *sandal wood* is a flame of fire." A careful reading of his papers, though, reveals a somewhat different story and an obstinacy that a mule might envy.

53 **publication of his results**   Brougham's two papers were "Experiments and Observations on the Inflection, Reflection, and Colours of Light (*Philosophical Transactions of the Royal Society of London* 86 [1796]: 227–77) and "Farther Experiments and Observations on the Affections and Properties of Light" (*Philosophical Transactions of the Royal Society of London* 87 [1797]: 352–85).

54 **registered as a medical student at Cambridge**   Young must have been quite a trial to his bemused tutors at Emmanuel College. The master introduced him by saying "I have brought you a pupil qualified to read lectures to his tutors," which can hardly have endeared him to them. He was a "fellow commoner" who dined with the fellows even though he was an undergraduate, did not attend lectures, and spent most of his time reading and doing experiments in his room. "[H]is room," said one of his fellow students, "had all the appearance of belonging to an idle man. I once found him blowing smoke through long tubes, and I afterward saw a representation of the effect in the *Transactions of the Royal Society*." Young's inability to communicate in comprehensible terms was well known to his fellow students. One wrote of him, "He was . . . worse calculated than any man I ever knew for the communication of knowledge."

54 **his London house**   A blue plaque marks the Georgian house at 48 Welbeck Street (near the more famous Baker Street) where Thomas Young practiced as a physician.

54 **the Bakerian lecture, "On the Mechanism of the Eye"**
Philosophical Transactions of the Royal Society of London 91
(1801): 23–88.

54 **the Royal Institution**   It rather pompously saw its role as
to bring together "the natural philosopher and those engaged
in arts and manufactures in order to improve industrial and
domestic efficiency."

54 **Young went about his task with characteristic earnestness**   Young delivered ninety-one lectures in the course of
his two-year stint. He resigned his position at the Royal Institution in July 1803, ostensibly because it was interfering with
his medical practice, but eventually he wrote up his lectures
(at the institution's request) in the form of two huge volumes,
with over a thousand pages of text and illustrations, and in
the expectation of substantial remuneration. He was to be
disappointed. By the time that the volumes appeared in 1807,
the publisher had gone bankrupt.

55 **"fatigued with insipidity or disgusted with inelegance"**
Cited in Geoffrey Cantor, "Thomas Young's Lectures at the
Royal Institution," *Notes and Records of the Royal Society* 25
(1970): 87–112.

55 **Young's second . . . Bakerian lecture**   "On the Theory
of Light and Colours" was published in *Philosophical Transactions of the Royal Society of London* 92 (1802): 12–48.

55 **claiming Newton's explanation of the Red River tides
as the stimulus for his theory**   A handwritten entry on
Young's notes for one of his Royal Institution lectures con-
firms this statement.

57 **interference colors**   These arise in some surprising circumstances. When I was a young scientist, studying the forces
between small particles suspended in liquids, someone found

a way to control those forces so that the particles were forced to organize themselves into a regular array, separated by distances close to the wavelength of light. When they did so, the normally rather drab white suspensions suddenly began to produce spectacular shows of sparkling interference colors. I showed a sample to my boss, an Englishman with a dry sense of humor who was not at all averse to putting down a brash young man from the colonies. He held the bottle up to the light, moved it from side to side, and then said with an expressionless face, "Hmmph. A phenomenon."

57   **an anonymous personal attack on Young**   Brougham has only recently been positively identified as the writer. His attacks on Young in the *Edinburgh Review* began in January 1803 (pages 450 and 457).

60   **a privately printed pamphlet**   Young's pamphlet was entitled *Reply to the Animadversions of the Edinburgh Reviewers on Some Papers Published in the Philosophical Transactions* (London: J. Johnson, 1804), a title that was not calculated to enhance sales.

60   **an experiment that anyone could try at home**   Young described the experiment in his third Bakerian lecture, "Experiments and Calculations Relative to Physical Optics," which was delivered on November 24, 1803, and published in *Philosophical Transactions of the Royal Society of London* 94 (1804): 1–16.

61   **"Dr. Young unwittingly stopped both portions; a thing extremely likely, where the hand had only one-thirtieth of an inch to move in"**   Experimenters who stretch the bounds of technique still seem to be regarded as fair game in some quarters. My colleague Jacob Israelachvili, who was the first person to control and measure the distance between solid surfaces to an accuracy of just one atom, once told me of the time when he announced his first results at an American conference. When he had finished, a very senior

academic came up to him and said that his own experience had shown that it was impossible to do such measurements, with the implication that therefore no one else could possibly have done them either.

62 **"the subject of Dr. Young's researches remained absolutely unnoticed"**   This was not entirely out of obstinacy and reverence for Newton. As the historian J. D. Mollon has pointed out, Young's theory could not at first explain the phenomenon of polarization, while Newton's theory at least had a fighting chance of doing so. J. D. Mollon, "The Origins of the Concept of Interference," *Philosophical Transactions of the Royal Society of London* A360 (2002): 816.

63 **Up until then everyone, including Young, had thought of the vibrations of light waves as being backward and forward in the same direction as the beam**   This includes Newton. Young's original inspiration was derived partly from Newton but also derived from his own early work on sound. In his reply to the *Edinburgh* reviewers, he says, "In making some experiments on the production of sounds I was so forcibly impressed with the resemblance of the phenomena that I saw, to those of the colors of thin plates, with which I was already acquainted, that I began to suspect a closer analogy between them than I could before have easily believed."

While he was at Cambridge, Young bet that he would "produce a pamphlet or paper on sound more satisfactory than anything that has already appeared, before he takes his Bachelor's degree." The *Parlour Book* of Emmanuel College for 1802 records that Young was adjudged to have lost the bet.

I avoided introducing the sound analogy earlier in this chapter because I could not find a way of doing so without overcomplicating the story.

63 **Augustin Fresnel**   While he was in prison in 1815 for fighting with the small army that tried to block Napoleon's return after his escape from Elba, Fresnel began to study

diffraction. Abstracts of his extensive original papers may be found in W. F. Magie, *A Source Book of Physics* (Cambridge, Mass.: Harvard University Press, 1965): 318–34. Fresnel died of tuberculosis on July 14, 1827, at the age of thirty-nine.

65 **he was still doing experiments designed to prove that Young was wrong**   Brougham's final paper, "Experiments and Observations upon the Properties of Light," was published in *Philosophical Transactions of the Royal Society of London* 140 (1850): 235–59.

66 **light . . . penetrates a tiny way into the air**   The light is called an *evanescent wave*. The experiment with the wineglass could, in principle, provide a much better way of recording fingerprints than coating the fingers with ink.

66 **The Modern Picture of Light**   This is described in Richard Feynman's wonderful little book *QED* (Princeton, N.J.: Princeton University Press, 1988). Unlike some other popular books by famous scientists, this one is eminently readable, and does not require mathematical training, only the willingness to work a little for the reward of understanding a complex and fascinating subject.

67 **The mathematics of wave propagation still works, and we still use it**   So great an authority as Robert Oppenheimer has said, "The work of . . . Fresnel on the wave properties of light is as necessary today as it ever was" (*Science and the Common Understanding*, BBC Reith Lectures [London: The Scientific Book Club, 1954]). Oppenheimer was dead right. As a scientist who regularly uses the phenomenon of interference to study closely spaced objects, I can vouch that hardly a day goes past where I do not use some aspect of Young's picture or Fresnel's mathematics. Oppenheimer's scientific career was destroyed in a witch hunt shortly after he delivered the Reith lectures. The hunter was Senator Joe

McCarthy, and the attack was on Oppenheimer's politics, not his science.

4. The Course of Lightning through a Corset

69 **People have always been fascinated by the power of lightning**   Even two of Santa's reindeer are named after thunder (Donner) and lightning (Blitzen). An excellent summary of the science of lightning and the history of belief in the power of lightning is given by Martin Uman in *The Lightning Discharge* (New York: Dover, 2001).

69 **It was especially drawn to Miss Minnie Frace**   I am pleased to report that Miss Frace survived the ordeal, as do most people who are struck by lightning.

69 **[Lightning] traveled down the steel ribs of the corset to blow her out of her shoes**   One journalist memorably commented that "sometimes the Lord moves in mysterious ways, and sometimes he comes straight to the point." http://www.njskylands.com/clcollege.htm.

69 **shoes, which are still preserved**   I am very much indebted to Jennifer Woodruff Tait, librarian for the Methodist Collection at Drew University in Madison, New Jersey, for tracking down the shoes and sending me several pictures.

70 **Benjamin Wilson**   A portrait painter of considerable repute, Wilson gained the ear of royalty when he succeeded William Hogarth as "Sergeant Painter to the King." Many of his portraits still hang in the National Portrait Gallery in London. His work *Earl and Countess of Derby, Their Infant Son and Chaplain* is on display at the Columbia Museum of Art in South Carolina.

70 **lightning rods**   A wonderful historical account of the lightning rod, with many fascinating anecdotes, is given in Richard Anderson, *Lightning Conductors: Their History, Nature and Mode of Application* (London: E. & F. N. Spon, 1885).

70 **The Greek playwright Aristophanes**   used the character of Socrates in *The Clouds* (420 B.C.) to describe the origin of lightning. Presumably the interpretation originally came from Socrates himself, but I have been unable to confirm this.

71 **five seconds corresponding to a mile in distance**   Or 3 seconds per kilometer. My father's formula was pretty accurate — just 2 percent out.

71 **some sixty people a year . . . receive injuries when lightning strikes the line while they are using the telephone**   "Telephone-mediated Lightning Injuries: An Australian Survey," *Journal of Trauma* 29, no. 5 (1989): 665–71.

71 **Franklin thought that the rubbing process transferred "electric fluid" from one material to another**   Franklin performed a classic experiment where he had two people standing on insulated blocks, with one holding a glass tube that he rubbed and then used to touch the other person. Each could then give a spark to a third person who was standing on the ground, but if they first touched each other, they lost the ability to deliver a spark. Franklin deduced from this that the action of rubbing the glass had given the first person a charge that was equal and opposite to the charge generated on the glass itself and passed over to the second person, and that when the two people touched, the charges canceled each other out. He then made the brilliant conceptual leap that the appearance of opposite charges was because one person had lost electric "fluid" and had hence become "negatively" charged, while the other person had gained it and become "positively" charged. Unfortunately he got the direction

wrong, which is why we now say that the material that moves (i.e., the electrons) has "negative" charge.

Some people take the idea of electricity "flowing" rather too literally. One who did this was the mother of the American humorist James Thurber, who would not leave a light socket empty in case the electricity dripped out.

76 **[Franklin] proposed an experiment that seemed to smack of human sacrifice**   He described it in a letter to his friend Peter Collinson on July 29, 1750. See Peter Collinson, *Experiments and Observations on Electricity Made at Philadelphia* (London: printed and sold by E. Cave at St. John's Gate, 1751), 65–66.

76 **Franklin records in his autobiography that his idea was "laughed at by the connoisseurs"**   Contemporary evidence seems to contradict this. R. V. Jones, "Benjamin Franklin," *Notes and Records of the Royal Society of London* 31 (1977): 201–25.

76 **a very French modification . . . of Franklin's suggested experimental arrangement**   See http://www.netrax.net/~rarebook/s980318.htm.

77 **his famous kite experiment**   Franklin reported it in *Philosophical Transactions of the Royal Society of London* 47 (1751–52): 565–67. The historian Tom Tucker argues in *Bolt of Fate* that Franklin never actually performed his experiment, but the degree of detail that Franklin passed on to Joseph Priestley and which Priestley later published in his *History and Present State of Electricity* makes this particular speculation hard to credit.

78 **an electrical corona forming around the tip of the shaft**   The coronas induced at the tips of the masts in old-time sailing ships by the strong electric fields that presage

lightning strikes were dubbed by sailors "St. Elmo's fire." Man-made fields can have equally striking effects; a fluorescent tube placed close to a high-voltage power line, for example, will begin to glow even though it is not connected to any external source of electricity.

78   **Under his enthusiastic promotion, pointed metal rods, connected electrically to the ground . . . were fitted to many public buildings**   The craze for "Franklin rods" continued for many years, eventually producing a plethora of door-to-door lightning rod salesmen, whose activities were satirized by Mark Twain in one of his funniest essays, *Political Economy:*

> Political economy is the basis of all good government. The wisest men of all ages have brought to bear on this subject the — [Here I was interrupted and informed that a stranger wished to see me at the door. I went and confronted him, and asked to know his business, struggling all the time to keep a tight rein on my seething political economy ideas. . . . He said he was sorry to disturb me, but as he was passing he noticed that I needed some lightning rods. . . . I try to appear (to strangers) to be an old housekeeper; consequently I said in an offhand way that I had been intending for some time to have six or eight lightning rods put up, but — The stranger started, and looked enquiringly at me, but I was serene. I thought that if I chanced to make any mistakes he would not catch me by my countenance. He said he would rather have my custom than any man's in town.

Twain goes on with his efforts to write his great essay on political economy, which are interrupted every couple of lines by the reappearance of the stranger at the door, anxious to sell him yet more lightning rods. Eventually his whole house becomes covered in lightning rods, to the wonder of his neighbors:

> For four-and-twenty hours our bristling premises were the talk of the town. . . . Our street was blocked day and night with spectators. . . . It was a blessed relief, on the second day, when a thunder-storm came up and the lightning began to "go for" my house, as the historian Josephus quaintly phrases it. It cleared the galleries, so to speak. In five minutes there was not a spectator within half a mile. . . . By actual count, the lightning struck at my establishment seven hundred and sixty-four times in forty minutes, but tripped on one of those faithful rods every time, and slid down the spiral twist and shot into the earth before it probably had time to be surprised at the way that the thing was done. And through all that bombardment only one patch of slates was ripped up, and that was because, for a single instant, the rods in the vicinity were transporting all the lightning they could possibly accommodate.

Twain's essay appeared a century after Wilson's first worries, but it seems to reflect them pretty accurately.

79 **lightning struck the Purfleet magazine**   The whole event is described with excruciating thoroughness in "Sundry Papers Relative to an Accident from Lightning at Purfleet, May 15, 1777," *Philosophical Transactions of the Royal Society of London* 68 (1778): 323–17. The magazine actually comprised five separate buildings. The only one still standing is the Purfleet No. 5 magazine, which is now a registered historical monument.

80 **London's grand Pantheon**   Described by Horace Walpole as "the most beautiful edifice in England," it was burned out in 1792, rebuilt in 1795, closed by the Lord Chamberlain

for holding theatrical performances without a license, converted to a bazaar, and eventually demolished to make way for the Marks and Spencer store that now occupies the site.

80   **The "lightning" came from a long tube**   Wilson's detailed description of his experiment, complete with scale diagrams and even photographs, is contained in "Sundry Papers Relative to an Accident from Lightning at Purfleet, May 15, 1777."

81   **His position was further undermined**   I am indebted to Professor Rod Home of the Department of History and Philosophy and Science at the University of Melbourne in Australia for providing me with an advance copy of his paper "Points or Knobs: Lightning Rods and the Basis of Decision-Making in Late-18th-Century British Science," presented at the Bakken Conference on the History and Cultural Meaning of the Lightning Rod (Minneapolis, Minnesota, November 4–6, 2002), in which he analyzes the lightning rod controversy in terms of "the nature and functioning of scientific authority and the way in which it may be enhanced or diminished." My information about public attendance at Wilson's experiments and Jean Hyacinth de Magellan's effect on Wilson's reputation was derived from Professor Home's article, and I am particularly indebted to him for his perceptive and useful comments about the content of this chapter from an historian's point of view.

81   **[Wilson] would not be the only scientist to have faked results that he needn't have faked**   I am referring here to actual faking, rather than selection of favorable results where, as Professor Rod Home has pointed out in a private communication, there are "some famous cases in the history of science where it seems pretty clear that not all unfavorable results got reported." I have discussed one of these cases (Millikan's oil drop experiment) in *How to Dunk a Doughnut*. There can be an argument for such selection, based on scien-

tific judgment of when an experiment has "gone right" or not, but there is no case for inventing or "improving" results to fit more closely with a theory. Fortunately, this seems to be extremely rare. I have had close contact with many scientists throughout my life and have known of only two cases where a scientist has "improved" results to provide more substantial support for what turned out to be a correct theory anyway.

82 **[King George III ordered] that Franklin's sharp-pointed lightning rods should either be removed from royal buildings or fitted with cannonballs** This prompted Franklin to write to him and say, "I have never entered into any controversy in defence of my philosophical opinions; I leave them to take their chance in the World."

83 **Michael Faraday . . . conceived the idea that all electrical charges produce an electric "field" through which they can exert a force on other charges** The notion of an electric field as something that can exert a force on an electrical charge has many useful applications. One is in radio and television reception, where the rapidly varying force that the field exerts on electrons in a metal aerial causes those electrons to vibrate in their turn, producing an oscillating electrical current that the receiver translates into sound and/or picture.

86 **Even if lightning hits the car, there will be no electric field *inside* the car** The car is an example of a "Faraday cage," a term that describes any space totally surrounded by a conducting material. The National Measurement Laboratory in Australia once built a whole laboratory, housing some 200 people, as a giant Faraday cage, with layers of copper mesh inside the double-brick walls and gold-coated windows to which the mesh was soldered. The effect was rather spoiled by all of the power cables that had to be brought in through the walls to power the instruments inside.

The poet Christopher Isherwood *almost* describes a Faraday cage in his poem "The Common Cormorant":

> The common cormorant or shag
> Lays eggs inside a paper bag
> You follow the idea no doubt
> It is to keep the lightning out.

It might have worked if the bags had been made of aluminum foil.

86 **The best rods seem to involve a compromise, with the tip slightly blunted**   Accurate computer calculations showed that the best results are achieved with rods whose "tip height–to–tip radius of curvature ratios" are about 680:1. C. B. Moore et al., *Journal of Applied Meteorology* 39 (2000): 593–609.

87 **Wilson argued that no one knew how much electricity a cloud contained**   Measurements have shown that all lightning bolts transfer pretty much the same quantity of electricity to earth (a few tens of coulombs), but this knowledge was not available to Wilson or his contemporaries, and his argument that pointed lightning rods posed an unquantifiable risk surely had some weight. Even the now-quantifiable risk can be pretty serious. The voltage difference between the ground and the base of the cloud is around 10 million volts, and the whole electrical charge is delivered by the bolt in a thousandth of a second. A quick calculation shows that the resulting current is around 10,000 amperes, which means that the power delivered by the stroke is 100,000 million watts — enough to light 1,000 million light-bulbs (if only for one-thousandth of a second!) and, as experience has shown, to shatter a tree. Even the enigmatic ball lightning carries a fair bit of power, as was shown when one ball contacted a large wharf pile and blew it into matchwood.

5. Fool's Gold

89 **According to the *Oxford English Dictionary,* the word [alchemist] derives from the ancient Greek** Others argue differently, claiming that the word first derived from the Arabic. Professor Lawrence Principe from Johns Hopkins University believes it probably derives from the Greek *cheo,* meaning "melt" or "smelt." I am happy to leave the resolution of the dispute to the philologists.

90 **Most alchemists were guided . . . by a belief that nature was resolvable into three principles** Boyle called these the *spagyrists,* who took their ideas from Paracelsus. Another, smaller group went all the way back to Aristotle for their basic belief, which was that all materials are composed of earth, air, fire, and water.

90 **We should be skeptical . . . of all statements concerning nature that are not based directly on experimental evidence** This paraphrase of Boyle's approach is given in the introduction to *The Sceptical Chymist* (Everyman's Library, 1967), x.

90 **The truth . . . uncovered by the American historian Lawrence Principe** The two ground-breaking books by Principe are *The Aspiring Adept: Robert Boyle and His Alchemical Quest* (Princeton, N.J.: Princeton University Press, 2000) and, with William R. Newman, *Alchemy Tried in the Fire: Starkey, Boyle, and the Fate of Helmontian Chymistry* (Chicago: University of Chicago Press, 2002). I have relied heavily on Principe's seminal work in the preparation of this chapter, and I am indebted to him for carefully checking and commenting on its content.

90 **alchemists were using quantitative experimentation**
Direct proof of this assertion comes from the famous picture
of a glass-cased balance in Elias Ashmole's 1652 *Theatrum
Chemicum Britannicum,* which is a virtual copy of the illumi-
nation from Norton's fifteenth-century manuscript *Ordinal of
Alchemy.* The associated glassware in the figure suggests
strongly that the balance was used for alchemy, and not simply for the assaying or weighing of gemstones.

90 **cardinals and others paid large amounts to have their
drinks cooled**   Newman and Principe, *Alchemy Tried in the
Fire,* 23–24.

91 **"a virtuoso uninfluenced by older ideas"**   Ibid., 33.

91 *The Alchymist in Search of the Philosophers' Stone Discovers
Phosphorus*   On exhibit at the Derby Museum and Art
Gallery, Derby, U.K.

91 **Socrates in the Louvre portrait**   The portrait (actually a
bust) dates from approximately 330 B.C.

91 **The alchemist was Henning Brandt**   Our knowledge of
Brandt and his work comes from the philosopher Gottfried
Wilhelm Leibnitz. Brandt wrote to Leibnitz with a description of his work, which Leibnitz later published in his *Histo-
ria Inventionis Phosphori* (Berlin, 1710).

91 **sixty buckets of human urine**   This was a favorite material of the alchemists, possibly because it was one of the few
readily available alkalis. The alkalinity comes from free ammonia, which is produced from the breakdown of urea in the
urine. The process does not take long to get started, as anyone who has ever changed a baby's diaper will know.

91 **white phosphorus**   One of several different physical forms,
or allotropes, of the element phosphorus, it was used in match-

heads because of the ease with which it would ignite, but its effects on the workers in that industry were devastating. Many ended up with "phossy jaw," which began with toothache and painful swelling of the gums and jaw. Abcesses then formed, accompanied by a fetid discharge which made its victims' presence almost unendurable. The victims' jawbones would literally rot and glow greenish white in the dark. The only treatment was for the jawbone to be removed surgically, an agonizing and disfiguring operation. White phosphorus was eventually outlawed in matches, being replaced by red phosphorus, an allotrope of phosphorus that does not ignite so easily but which does not cause "phossy jaw."

92 **glowed when first exposed to air**   The glow was what we would now call *chemiluminescence,* which is energy released in the form of light by a chemical reaction. The reaction in this case was that of phosphorus with traces of oxygen in the retort. Some chemiluminescent reactions, such as those in the tail of a firefly, can produce light without heat, but the reaction of phosphorus with oxygen produces large quantities of heat, and when Brand let air into the flask, the phosphorus burst into flame, filling the room with choking fumes of poisonous phosphorus pentoxide.

92 **"If the Privy Parts be therewith rubb'd . . . they will be inflamed"**   White phosphorus dissolves in oily materials, and when it is placed on the skin it penetrates and dissolves in the subcutaneous fat, from where it can migrate to cause terrible damage to kidneys, liver, etc. Definitely not to be trifled with!

92 **Brandt's experiment seems foolhardy and ridiculous to modern eyes**   One might contrast his approach with that of the 1930s, when radioactive materials were believed to be health-giving on the basis of no experimental evidence whatsoever, and where spa waters that were loaded with radium fetched particularly high prices on that account! Unbelievably, some modern spas are still claiming health benefits

from radium in spa waters. No, I'm not going to say which they are. Look on the Internet. There is at least one in Korea, and another in Germany.

93 **"vitriolification"** A listing of this and other processes used by the alchemists is given in a remarkable website — http://www.levity.com/alchemy/index.html — which contains a huge amount of well-researched information about all things alchemical.

93 **kept in the family medicine cabinet in case of snakebite** If any of us had been bitten by a poisonous snake (of which there were quite a few around!), the Condy's crystals would have been totally useless, as would the prevailing advice to slash the wound with a razor blade and suck out the poison. The best way to treat a bite is to place a tourniquet above the wound and get the patient to a doctor pronto.

94 **calomel [was] sometimes still prescribed as a purgative when I was a child** The use of calomel (mercurous chloride) as a purgative was not outlawed in New South Wales until 1957. It was a major cause of Pink disease, an often fatal condition caused by mercury poisoning.

95 **experiments in which he took "a *considerable quantity* of *Man's Urine*"** It might be noted in passing that Boyle performed most of his experiments in his own home. It is little wonder that he never married.

95 **The charcoal block and blowpipe were still in regular use . . . when I began to study chemistry** William R. Newman makes a case for the very early use of the blowpipe by alchemists — well before its introduction into "real" chemistry by Johann Kunckel in the seventeenth century. William R. Newman, "Alchemy, Assaying, and Experiment," in *Instruments and Experimentation in the History of Chemistry,*

ed. Trevor H. Levere and Frederic L. Holmes (Cambridge, Mass.: MIT Press, 2000), 35–54.

98 **Glauber's Salts** Or sodium sulfate; its first use was in the dye industry, but it was subsequently found to be an effective laxative (it would be interesting to know how this discovery was made). It has been carried for this purpose in the medicine chest of many an exploratory expedition.

98 **The crystals would produce fantastic growths** The growths consist of insoluble silicates of copper, cobalt, etc.

99 **"Metalls grow into lovely trees"** No, they don't. The starting material needs to be a soluble salt, not an insoluble metal.

100 **very small particles that he called "corpuscles"** The existence of corpuscles was propounded by Boyle in his *Essay on the Atomicall Philosophy* (Newman and Principe, *Alchemy Tried in the Fire*, 19). One of Boyle's most important experiments to prove the existence of corpuscles was to dissolve a weighed amount of silver in nitric acid, filter the solution, and recover the silver by precipitating it with salt of tartar (i.e., potassium carbonate obtained by heating potassium bitartrate). He showed that the amount of silver recovered was just the same as the amount that he had started with, and so concluded that the silver had not been destroyed when it was dissolved in the acid, but only broken up into corpuscles too small to be seen. The small size of the corpuscles was confirmed by the fact that they could pass through a very fine filter. It was a clever experiment, but Boyle was not its originator; he had lifted the idea in toto from the seventeenth-century alchemist Daniel Sennert, who had designed and performed the experiment to prove the very point that Boyle claimed as his own.

101 **"Philosophical mercury"**   The concept is virtually impossible to understand in modern-day terms. According to Newman and Principe, "The basis of the process [for preparing 'philosophical mercury'] lies in treating common mercury with the martial regulus of antimony in order to 'ennoble' or 'actuate' it [into a] metallic solvent capable of 'radically' dissolving gold into its principles and readying it for preparation into the Stone." Ibid., 120.

102 **[Boyle had] "made trial of our Mercury"**   "Of the Incalescence of Quicksilver with Gold, Generously Imparted by B.R.," *Philosophical Transactions [of the Royal Society of London]* 10 (1675): 515–33.

106 **"universal solvent"**   Its composition was similar to that of *aqua regia,* which is a mixture of nitric and hydrochloric acids.

106 **a demonstration of the stone in action**   The incident is described in Principe, *The Aspiring Adept,* 98.

107 **scientists are as gullible as anyone else**   That includes me. I remember an occasion when I was trying to grow mushrooms in a box in a dark cupboard, and looked every day to see if any had appeared. My wife bought a bag of tiny mushrooms one evening, and secretly thrust their cut ends into the surface of the compost. When I came to look the next morning, the surface of the compost appeared to be covered with freshly grown mushrooms. I should have realized that such a rate of growth was highly unlikely, but I didn't stop to question how the mushrooms had managed to appear so fast, and it was only after I had rung several friends to tell them of this remarkable phenomenon that my wife revealed the truth.

107 **repeal of the three-hundred-year-old Act Against Multipliers**   According to the Royal Society Journal Book, "Inducement to the Parliament to Repeal the Act of Henery 4th

against Multiplication of Metalls was from the testimony of Mr. Boyle and the Bishop of Salisbury who affirm to have seen projection, or the transmutation of other Metalls into Gold."

108 **turn gold atoms into mercury atoms** *New Scientist,* August 23, 2003.

6. Frankenstein Lives

Nobody who writes about the Galvani-Volta controversy can do so without reference to Marcello Pera's very detailed and authoritative account, *The Ambiguous Frog,* translated by Jonathon Mandelbaum (Princeton, N.J.: Princeton University Press, 1992). The approach that I have taken in this chapter is my own, but I acknowledge here my extensive debt to Pera's work (hereinafter referred to as *TAF*), and I have tried to indicate as accurately as possible those places where I have quoted from it directly.

111 **Movie audiences of the early 1930s** Universal Studios produced *Frankenstein* in 1931. It bears little resemblance to Mary Shelley's book, and even gets Frankenstein's name wrong (it is Victor in the book). Karloff's name is changed, too, for what was probably a better reason; in real life he was William Henry Pratt.

111 **Colin Clive . . . used a high-voltage surge of electricity** Mary Shelley, with a better-developed sense of mystery, did not actually reveal in the book the method by which the monster was brought to life.

111 **Aldini's London show** took place at London's Royal College of Surgeons on January 17, 1803. Forster had been hanged at Newgate only an hour before. One of the conditions of his sentence was that his body should be used for "medical research," although the judge can hardly have had

Aldini's demonstration in mind. According to an eyewitness, "Galvanism was communicated by means of three troughs combined together, each of which contained forty plates of zinc, and as many of copper. On the first application of the arcs the jaw began to quiver, the adjoining muscles were horribly contorted, and the left eye actually opened." As the experiment proceeded, Forster's clenched fist rose in the air, his legs started to kick violently, and his back arched. Not a pretty sight.

A full scientific account of Aldini's experiment is given in *Journal of Natural Philosophy* 3 (1802): 298–300 and in *Philosophical Magazine* 14 (1802): 364–68.

112 **The audience at an experiment in Glasgow scattered wildly** The experimenter was Andrew Ure. The experiment took place in 1818, and the corpse was that of the hanged thirty-year-old murderer Matthew Clydesdale, which responded to electric stimulation by kicking out a leg with such violence as to nearly overturn an assistant who was vainly trying to hold it down. Ure later came very close to claiming in print that the body had come back to life.

112 **Aldini's demonstration was one of the inspirations for Mary Shelley's book** Mary Shelley was only six at the time, but it seems very likely that she heard accounts of the event from firsthand witnesses in whose circles she moved. *Chemistry in Britain,* November 2002, 26.

112 **"magical picture"** *TAF,* 14–15.

113 **he was not endowed "with everything that constitutes the distinctive character of a man"** *TAF,* 16.

113 **"men who have been mutilated by Art and men towards whom Nature has displayed cruelty"** *TAF,* 17.

114 **one doctor in Montpellier** This was Jean-Etienne Deshais, *TAF,* 22.

116 **Luigi Galvani . . . decided to test whether electricity might also have an effect on the limbs of dead animals** Galvani seems to have begun his experiments in 1780 and published the results in 1792. My description of his experiments largely follows that of *TAF*, although I have checked other sources for extra detail where possible.

118 **It seemed that it was the spark itself, and not the current, that caused the frog's muscles to contract** Nowadays we would say that his wife, in discharging the generator, had also discharged the frog that was attached to it, with the resultant flow of electrical current causing the legs to twitch. This was in fact what Volta said later, but Galvani took a different tack.

122 *De Viribus Electricitatis in Motu Musculari Commentarius* Galvani actually sent a copy in March 1792 to Don Bassiano Carminati, professor of medicine at Pavia and close friend of the leading Italian physicist, Alessandro Volta, who also worked in Pavia. Galvani clearly hoped that Carminati would talk to Volta and obtain his approval, but it didn't work out quite like that.

123 **"The storm aroused by the publication of the commentaries"** The quote is from Emil Du Bois-Raymond, *Untersuchungen über thierische Elektricität* (Berlin: Reiner, 1948); quoted in *TAF*, 117.

124 **"contact potential"** You can experience the effect of a contact potential for yourself by biting down on a piece of aluminum foil with a gold or gold-amalgam-filled tooth. The contact between the aluminum and the metal in the tooth will produce quite an unpleasant sensation, due to electrical stimulation of the tastebuds.

124 **the invention of the battery** It may be that Volta was beaten to the invention of the battery by some two thousand

years. Some fifty years ago, Wilhelm Konig, then director of the Baghdad Museum, discovered a small pottery jar in the ruins of ancient Parthian settlements at Seleucia, near Baghdad, that had all of the characteristics of a battery. Inside the three-centimeter opening at the top there was a copper tube held in place with asphalt. The tube was sealed in place at the bottom with a copper disc held in place by more asphalt. Suspended from the asphalt lid was an iron rod that hung down the axis of the copper tube.

Such a jar filled with vinegar would have acted as a battery. According to Paul T. Keyser of the University of Alberta ("The Purpose of the Parthian Galvanic Cells: A First-Century A.D. Electric Battery Used for Analgesia," *Journal of Near Eastern Studies* 52 [April 1993]: 81–98), a series of such batteries may have been used in the treatment of pain. Bronze and iron needles found nearby suggest that they may even have been used in acupuncture.

125 **"Electromotive apparatus contained in the fish"**   The electric organ contained in the tail of the electric eel does indeed resemble a series of stacked plates that together can produce an enormous, and sometimes fatal, jolt. The manner in which they produce that jolt, however, is very different from the way in which a battery produces electricity. The plates of a battery produce their electricity through a chemical reaction that slowly but surely dissolves them or changes their chemical composition — a process that Volta might have noticed but didn't. With some batteries, the process can be reversed (this is what happens when you "charge" a battery), but only by putting more electricity back in than was taken out in the first place. In time, the plates of the battery degrade and the battery becomes useless.

127 **setting up a short circuit**   In the text I have short-circuited Peter's full explanation, which gives more detail of what happens when a wire is pushed through an axon:

short circuiting them in this damaged region, and making an electrical connection with the internal solution within the nerve cells. When the other end of the wire was electrically connected to tissue and thereby to fluids bathing the intact region of the nerve near the attached muscle, this completed a circuit that induced the nerve to initiate an electrical signal (an action potential) at that point, which would travel down to the junction of the nerve and muscle (the end-plate), cause it to release neurotransmitter, which would diffuse to the muscle cell membrane, electrically excite it and, in turn, cause the muscle to contract.

127 **how an action potential is propagated along the axon**
The evocative term *action potential* was coined by the German physiologist Julius Bernstein around 1870, but it was not until the 1950s that the Cambridge physiologists Alan Hodgkin and Andrew Huxley were able to describe and predict its behavior mathematically. I was lucky enough, some thirty years later, to spend a year working in the laboratory where the discovery was made and to learn a little of how such discoveries come about. Huxley was no longer there, but Sir Alan (as he then was) still worked in an office whose floor he had refused to have carpeted because he considered it an unnecessary expense. I only met him once but even then was impressed by the total focus that he brought to a problem. It was an approach that permeated the laboratory, but the real difference between his and many other laboratories that I have known was that the focus was on *important* problems rather than problems that would produce guaranteed results.

Hodgkin and Huxley's work (which led to the award of the 1963 Nobel Prize for Physiology or Medicine) produced a spate of experiments aimed at elucidating the underlying molecular mechanism, which relies on a bevy of channels and pumps that span the membrane.

When the neuron's job is to stimulate a muscle (i.e., a

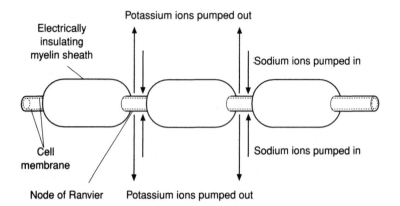

Figure 6.4. Schematic representation of a giant squid axon in action ("giant" refers to the axon, not the squid!)

"motor" neuron), the axon is electrically insulated by a sheath of protein (myelin), which is interrupted every two millimeters or so by a very short gap called a Node of Ranvier, so that the whole thing looks like a miniature string of sausages. These nodes are the sole places where ion transfer takes place.

In its "resting" state the charges across the membrane are unbalanced, with more sodium ions on the outside and more potassium ions on the inside, producing a voltage of some 65 millivolts across the membrane. When the neuron receives a chemical signal from an adjacent neuron, it sets off a chain of events that is repeated for each Node of Ranvier as the nerve impulse "jumps" down the axon (this is known as "saltatory" conduction). The first step is that nearby sodium channels are stimulated to open, allowing positively charged sodium ions to flood in. This causes a further rapid change in voltage (depolarization), which reaches a peak that stimulates potassium channels to open, venting potassium ions to the outside and restoring the initial voltage conditions (in fact, there is a slight overshoot). This pulse of depolarization travels along the axon to the next node, where the process is repeated. In the meantime, an active "pump" uses energy that ultimately comes from the food you eat to shift ions back to where they came from and restore the conditions that existed before firing. The whole process takes about a millisecond, after which the nerve is ready to produce another pulse.

Alcohol intoxication produces some of its behavioral side effects by affecting the membrane potential. Ethanol metabolism produces fatty acid ethyl esters, which can hold the potassium channels open, allowing potassium ions to leave the neuron even when it is in a resting state. This makes it harder for the membrane to reach the potential necessary to initiate an impulse, resulting in impairment of mental and motor functions. Excessive amounts of potassium in the diet can cause the same effect, which can lead to death from nervous system failure.

130 **"there is no evidence that TENS provides effective pain relief"**   See the Oxford Pain Internet Site, http://www.jr2.ox.ac.uk/bandolier/booth/painpag/.

130 **the case of a Swiss locksmith named Nogues**   See Ellen R. Kuhfeld, "Proceedings of the Fourth International Symposium on Biologically Closed Electric Circuits," Bloomington, Minnesota, October 26–29, 1997. This article is published on the Internet at http://members.aol.co kuhfeld/Return.htm under the title "Return with Us Now — To Those Shocking Days of Yesteryear."

7. What Is Life?

I am grateful to Jane Maienschein, Regents' Professor of Philosophy and Biology at Arizona State University and an expert on the history of cloning and the Driesch-Roux controversy, for her helpful comments on the content of this chapter. The story of the Driesch-Roux controversy is only one strand of the rich history of developmental mechanics, or *Entwicklungsmechanik*. For more fascinating detail, see Jane Maienschein, "The Origins of

Entwicklungsmechanik," in *A Conceptual History of Modern Embryology*, ed. F. Scott Gilbert (Baltimore: Johns Hopkins University Press, 1994), 43–61.

133 **an experiment to manufacture an animal clone**   A clone is, technically, a set of genetically identical mature organisms. There are many ways of producing such a set, all of which could legitimately be described by the word *cloning*. Professor Maienschein has written extensively on this subject (see, for example, http://bioethics.net/hottopics/stemcells/ excerpts_ Maienschein.php) and has argued in private correspondence with me that the word *cloning*, as it is presently used in popular parlance, implies that nuclear transfer has taken place — i.e., the nucleus of one cell has been transferred to a second, enucleated cell, which is then allowed to develop by division and differentiation into a mature organism. This process permits the production of animal offspring that are genetically identical to the parent, as was the case with Dolly, and which could controversially be the case with human beings.

Professor Maienschein's point is an especially important one in view of the legal issues that now surround the subjects of animal cloning in general and human cloning in particular. In talking about the scientific issues, therefore, I have avoided the word *cloning* in favor of the circumlocution "production/manufacture of a clone" to encompass all methods of producing genetically identical organisms from cells other than the original fertilized egg.

133 **Even as late as 1930 the eminent historian Emmanuel Radl was able to write that "Driesch marks the end of Darwinism"**   E. Radl, *The History of Biological Theories* (Oxford: University Press, 1930): 352. Quoted in L. F. Koch, "Vitalistic-Mechanistic Controversy," *Scientific Monthly* 85 (1957): 245–55.

134 **It was Aristotle himself who wrote the first book on the subject**   The idea of vitalism did not originate with

Driesch, as Driesch himself would have been the first to acknowledge. In his 1914 book *The History and Theory of Vitalism,* he traces the idea back to Aristotle, who wrote in his
book *De anima* that the mind or soul is the "first entelechy"
of the body, the "cause and principle" of the body, the real-
ization of the body.

135 **the Dutch microscopist Anton van Leeuwenhoek reported seeing homunculi in the heads of living spermatozoa** Leeuwenhoek was a tradesman in Delft who
turned from making microscopes to using them when his
friend and fellow townsman, the prominent young anatomist
Regnier de Graaf, brought him to the attention of the
Royal Society in London as a maker of exceptional microscopes. It was actually a student from the medical school
at Leiden, Johan Ham, who brought him a specimen of semen in which Ham had observed small animals with tails,
which Leeuwenhoek now observed as well. Leeuwenhoek described these observations in a brief letter to Lord Brouncker,
then president of the Royal Society, but had considerable
doubts about the propriety of talking publicly about semen,
and begged Brouncker not to publish if he had any doubts on
this score. For more detail about this fascinating episode in biological history, see http://www.devbio.com/article.php?ch=
7&id=65.

135 *The Ovary of Eve* Clara Pinto-Correia, *The Ovary of Eve*
(Chicago: University of Chicago Press, 1997).

136 **urea, an "organic" substance . . . could also be manufactured from materials that were in no sense living**
This could be done by reacting cyanic acid with ammonia. F.
Wöhler, *J. C. Poggendorff's Annalen der Physik und Chemie* 88
(1828): 253–56.

138 **He conceived the brilliant experiment** W. Roux,
"Beiträzur Entwickelungsmechanik des Embryo," translated

in *Foundations of Experimental Embryology,* ed. B. Willier and
J. M. Oppenheimer (New York: Hafner Press, 1974), 2–37.

138 **the International Zoological Laboratory [at] Naples**
Coincidentally, this was that place where Jim Watson gained
the inspiration from a talk by Maurice Wilkins that led to
his solving (with Francis Crick) the structure of DNA and to
the award of a Nobel Prize to all three scientists. It is interesting to speculate what might have happened if Rosalind
Franklin, on whose data the Watson-Crick interpretation was
based, had still been alive when the prize was awarded, since
joint Nobel Prizes are never awarded to more than three
people. This is one of the numerous failings of such prizes,
which do not take account of the fact that science is a community activity where individual contributions are often hard
to isolate.

138 **Driesch carefully separated the daughter cells of a sea
urchin embryo**   H. Driesch, "Entwicklungsmechanische
Studien I," translated and abridged in Willier and Oppenheimer, eds., *Foundations of Experimental Embryology,* 38–50.

139 **"How could a machine be divided . . . and yet remain
what it was?"**   Hans Driesch, *The History and Theory of Vitalism,* trans. C. K. Ogden (London: MacMillan & Co., 1914),
211–12.

139 **The mechanists . . . [adopted] an extraordinary hubris**
My personal award for such hubris goes to Carl Vogt, the first
rector of the University of Geneva, who stated categorically
that the brain produces thoughts as physical materials in the
same way that the kidneys produce urine and the liver produces bile. It is true that nerve conduction is associated with
the secretion of specific chemicals (called neurotransmitters),
but Vogt had something far more direct in mind, as do some
modern philosophers. When I began to study philosophy for-
mally late in life, I discovered that some modern philosophers

conceive of "thoughts" in a way that seemed to me remarkably like that of Vogt. One popular argument, called the representational theory of mind, suggested that "thoughts" are equivalent to physical structures in the brain. This seemed to me either to be such a trivial statement as not to be worth making, or to be so profound (it took a whole book to explain it) that a mere scientist like myself had no chance of understanding it.

140 **Hans Spemann . . . discovered that the development of the living daughter cell**   See Klaus Sander and Peter Faessler, "Introducing the Spemann-Mangold Organizer: Experiments and Insights That Generated a Key Concept in Developmental Biology," *International Journal of Developmental Biology* 45 (2001): 1–11.

140 *What Is Life?*   Erwin Schrödinger, *What Is Life?* (Cambridge: Cambridge University Press, 1944).

140 **wave mechanics, in which a single equation is used**   The equation in its most compact form is just $H\Psi = E\Psi$. These symbols conceal a wealth of complexity, including the famous Schrödinger's cat paradox, which makes another surprising link between physics and biology. The paradox comes about because wave mechanics (which is one description of quantum mechanics) predicts that atoms and other small particles can be in two states at once (e.g., spinning clockwise and counterclockwise simultaneously) and that it is only when an observer looks that these particles definitely enter one state or the other. Schrödinger emphasized the peculiarity of the paradox in 1935 by applying the same logic to a cat hidden in a shut box along with a radioactive substance, from which the decay of just one atom would lead to a chain of events that opened a vial of poison and killed the cat. Until the box is opened, an observer has no way of knowing whether or not the atom has decayed, so by quantum mechanical arguments the atom must be in both states — decayed and non-

decayed — until the observer opens the box and looks, where it enters one state or the other. By the same logic, the cat must be both dead and not-dead until the observer looks! Schrödinger is supposed to have said later in life that he wished he'd never heard of the cat, which has surely provoked more discussion than any other topic in quantum mechanics. It has also provoked experiments that have shown that Schrödinger's description was right in principle, e.g., J. R. Friendman et al., *Nature* 406 (2000). In these experiments, the "dead and alive" cat was replaced by an electric current flowing around a ring both clockwise and counterclockwise.

Stories about Schrödinger and his carelessness and unconventional behavior abound. According to Robert Weber in *Pioneers of Science,* when Schrödinger went to a conference "he would walk from the station to the hotel where the delegates stayed, carrying all his luggage in a rucksack and looking so like a tramp that I needed a deal of argument at the reception desk before he could claim a room."

141 **Schrödinger's book stimulated Francis Crick to move from physics to biology**   See J. D. Watson, *The Double Helix* (New York: Atheneum Press, 1968).

145 **Melzer's analysis**   S. J. Melzer, "Vitalism and Mechanism in Biology and Medicine," *Science* 19 (1904): 18–22.

146 **different types of cell might have different "adhesivities"**   Malcolm Steinberg, "Reconstruction of Tissues by Dissociated Cells," *Science* 141 (1963): 401–8.

147 **Richard Owen**   Sir Richard Owen (1804–1892) was a pioneering British comparative anatomist who coined the term *dinosauria* (from the Greek *deinos,* meaning "fearfully great," and *sauros,* meaning "lizard"), recognizing dinosaurs as a suborder of large, extinct reptiles.

Owen entered the University of Edinburgh medical school in 1824. However, he was displeased with the quality of teaching, especially in comparative anatomy. Like Darwin after him, Owen enrolled in Barclay School, a private school offering instruction in anatomy. Here he was deeply influenced by John Barclay, who was an avowed antimaterialist. At the time, Edinburgh physicians and scientists were hotly debating whether mind and life could be reduced to material explanations, or whether mind was a completely separate, nonphysical entity that could not be reduced to physical phenomena. Barclay supported this antimaterialist, dualist view, arguing that the essence of life was a "Vital Principle" and the essence of mind was a "Soul," neither of which was material. Owen was strongly influenced by Barclay's views.

148 **My dear Julian, I never could make sure about that water baby**   Quoted in Julian Huxley, *Memories* (London: George Allen & Unwin, 1970), 24–25.

150 **Phil Vardy, Keith's anarchic Ph.D. student**   Phil was pursuing the clues that lay in the slime trails left by moving *grexes* as they traversed the surfaces of the petri dishes that littered his laboratory bench. The trails were not like those of snails or slugs but were collapsed tubes of material that the *grex* had secreted around itself and which it had now left behind. The trails contained "footprints" — a series of oval- and crescent-shaped marks that were invisible in ordinary light but showed up as bright green patches under the microscope after they had been selectively stained with a fluorescent dye. Phil was trying to correlate these marks with the peculiar shuffling motion of the *grex* as it alternately shortened and lengthened its body. Phil photographed the trails through a powerful microscope, matching up the overlapping photographs to produce a composite photograph that was some ten *meters* long, and which could be correlated with the actual movement of the *grex* to show that the "footprints" were really "noseprints." Phil hypothesised that the *grex* pulled itself along by first extending its front end in the manner of an arching proboscis, gluing the tip of the proboscis to the surface and shortening it to pull

the rest of the "body" along, before detaching the tip to repeat the process. It was a beautiful piece of work, duly rewarded by appearing as an article in the prestigious scientific magazine *Nature*. It was here that my ideas as a physical scientist finally reaped their reward, and I was lucky enough to have my name attached to the paper after I used a specialized type of microscopy (called polarizing microscopy) to reveal stressed areas in the sheath and hence show that the "nose" really was generating pulling forces.

152 **Lewis Wolpert**   Mathematicians now know models in which the developmental fate of a cell is determined by its position in a set of chemical gradients as "Wolpertian."

152 **Alan Turing**   Much has been written about Turing and the way that he was hounded by the establishment on account of his homosexuality. Less well known are his exploits as an athlete and games player, which he combined in the invention of "round-the-house" chess, where a player has to get up and run out the door and around the house before being allowed to make his or her next move. If the runner gets back before the opponent has made a move, he or she is entitled to make another move (and then run around the house again). I tried this game once. Never again.

155 **Spemann's experiments caused huge public interest**   See K. Sander and P. E. Faessler, *International Journal of Developmental Biology* 45 (2001): 1–11.

155 **She had discovered the "organizer region"**   A number of different "organizer regions" have now been discovered. They work by releasing chemicals at concentrations that affect the developmental paths of their nearest neighbors.

157 **"purely local interactions can generate arbitrarily complex spatio-temporal patterns"**   See, for example, Conway's "Game of Life," http://www.brunel.ac.uk/depts/AI/alife/al-gamel.htm.

Index

Page numbers in italics indicate illustrations.